Ancient Skies and Astronomy Now

Krishna Ramadas

ISBN-13: 978-1493510313, ISBN-10: 1493510312

This book is dedicated to my Guru Sri Sri Ravishankar and the system of Gurukul that has preserved the Vedic knowledge through the millennia.

Contents

Preface

Over the past few decades, astronomers have made huge strides in the formulation and validation of new theories of the creation and evolution of the cosmos. Precision equipment scans the radiation sources of deep space in order to refine our conception of the formation of stellar systems and the structure of our own Milky Way Galaxy.

In ancient times, especially during the period of the Vedic civilization, oral tradition and memorization fundamentally underlied all education. Vedic approaches to learning and preserving knowledge challenged the human brain in ways that modern researchers of neuroplasticity can appreciate.

Striking similarities between ideas of ancient astronomy, expressed in the language of Vedic mantras, and modern astronomy raise intriguing questions about the sources of ancient knowledge. Was the human brain, sharpened through techniques of meditation and yoga designed to increase its plasticity, capable of far reaching intuitive discoveries? Vedic Rishis have attributed their discoveries to a process of cognition that arose in deeper states of expanded meditation and consciousness. Their descriptions of myriad aspects of the ancient skies resemble many discoveries of modern astronomy. Modern astronomy software enables anyone to roll back time today to look at the configuration of the skies millennia ago, to see the stars as the Vedic Rishis of yore once saw them.

I would like to thank my wife Suchitra and our sons Rohan and Vikram for their editorial help and for encouraging me to write this book. I hope this book will kindle interest in its unique presentation of astronomical discoveries, far reaching and brow-raising in their accuracy, of the Vedic texts.

Krishna Ramadas

California, October 2013

1. *The Continuing Creation*

In their own unique style, the Vedic Rishis attributed the origin of the universe to an intention arising in the cosmic mind, which then gave rise to the primordial sound. Modern science's description of the creation of the universe reads like the words of the Rishis, like a chapter from the Vedas. The universe we see today evolved from a dense primordial plasma which contained three entities: dark matter, baryons and photons. The densely packed material of the plasma generated a powerful gravitational pull inwards. Photons and matter collided in the tight space, releasing heat that created outward pressure. In the midst of the counteracting forces, a perturbation arose deep within the plasma, upsetting its delicate balance.

The perturbation set the plasma oscillating, not entirely dissimilar to sound waves created in still air due to sudden differences in pressure. This oscillation resulted in a spherical sound wave (C. L. Bennett) moving from the depths of the plasma towards its edge, carrying with it baryons moving over half the speed of light. The dark matter components of the plasma remained stationary, unaffected by sound waves while photons and baryons progressed beyond the edge of the primordial plasma. This, in a nutshell, was the birth of our universe.

The nascent universe, dense in the material of the plasma, did not permit photons to travel freely and was thus enveloped in darkness. As the plasma began to cool, electrons and protons began to combine and coalesce into neutral hydrogen atoms. Over the next half a million years, with the plasma continuing to expand, the photons were at last freed from the neutral environment and streamed outwards to the visible edges of today's universe.

Using modern instrumentation, these primordial photons can be detected today all throughout the deep sky as the Cosmic Microwave Background (CMB). Figure 1 shows the structure of the universe formed by the CMB radiation, the energy that escaped at the initial big bang. Clearly visible is the contrast between the structure formed by the invisible CMB radiation and the visible structure formed by what the escaping energy left behind.

Towards the edges of the region where energy escaped the universal plasma, a shell formed at a distance now referred to as the sound horizon (Chicago, 2013). With this energy gone, gravitational forces began to dominate, producing configurations of high baryon densities. The matter within these structures gradually dispersed and consolidated to form galaxies over billions of years of time. A greater number of galaxies are seen to be separated by the sound horizon than by other time distances because of the initial configuration of matter.

The *Nasadiya Sukta* of the Rig Veda describes the beginning of the universe thusly:

> *Nasadasinno sadasittadanim nasidrajo no vyom paro yat*
>
> *Kimavarivah kuha kasya sharmannambhah kimasid gahanam gabhiram*

Beyond space, in the beginning, there was neither existence nor non-existence; neither birth nor death. All was energy. The Nasadiya Sukta goes on to offer a multitude of phrases -- "darkness that wrapped around darkness' – that, without the depth of understanding provided by modern cosmology, make ostensibly little sense.

The translation of the Nasadiya Sukta in its entirety is provided here:

> *At first was neither Being nor Non-being. There was not air nor yet sky beyond. What was it wrapping? Where? In what protection?*
>
> *Was water there, unfathomable and deep?*
>
> *There was no death then, nor yet deathlessness; of night or day, there was not any sign.*
>
> *The One breathed without breath, by its own impulse. Other than that was nothing else at all.*
>
> *Darkness was there all wrapped around by darkness, and all was water indiscriminate. All was energy indiscriminate.*
>
> *Then that which was hidden by the void, the one emerging, stirring, through power of Ardor, came to be.*
>
> *In the beginning Love arose, which was the primal germ cell of the mind.*
>
> *The seers, searching in their heart with wisdom, discovered the connection of the Being in Non-being. A crosswise line cut the Being from the Non-being.*
>
> *What was described above it, what below? Bearers of seed there were and mighty forces, thrust from below and forward move above.*

Who really knows? Who can presume to tell it?

When was it born? Whence issued this creation? Even the Gods came after its emergence.

Then who can tell from whence it came to be?

That out of which creation has arisen, whether it held it from or it did not; He who surveys it in the highest heaven, He surely knows or maybe He does not!

In the Vedic tradition, knowingness is inherent to the cosmos. Nature and the instincts therein are but a segment of this idea. A flower blooming, a bird constructing its nest, and the human intellect inventing tools and machinery are all a part of this knowingness. While atoms and molecules assemble to create the human form, a part of the cosmic knowingness crystalizes into human intellect. And it, the knower, continues to be an integral part of the cosmos. The knower itself may be ignorant or otherwise, according to the Naasadiya Sukta: "He surely knows or maybe He does not."

Space contained the primordial soup and all its constituents before the big bang, and it holds the universe today. Vedic tradition anthropomorphizes this principle as *Brahman*. The *Brahman* of the Nasadiya Sukta -- *Parame Vyoman* -- is the space that contains, pervades, and transcends the light of the stars, the material of the galaxies, and all the elements of creation.

Galaxy Formation:

The nascent universe consisted of newly formed neutral hydrogen and was characterized by variations in density. Over time, some areas increased in density, were dispersed, and became identifiable as distinct amalgamations of matter. The densest regions approached structural collapse, generally smaller in size. Others continued to grow more slowly and fragmented upon collapse. These initial formations morphed into proto-galaxies, which started appearing about 14 billion years ago. Figure 2 show the formation of a proto galaxy from a lump of gas and dark matter.

A galaxy is one of cosmology's more fundamental units. Incorporating stars, gas, dust and dark matter, the successful formation of a galaxy includes the generation of millions of stars. The physics of galaxy formation is complex, marrying theories of thermodynamic, stellar energy production, and dynamic gravitational interactions. Stars can form from clouds of gas, but a new star can also dissipate clouds by heating them, thus preventing the formation of other stars. Partially a result of this fundamental formative process, the space between galaxies ranges on the incomprehensible. The galaxies we see in the sky today, like the fossilized remains of ancient species, provide clues to times long past.

A proto-galaxy starts its life as a huge expanse of gas that gradually develops into clouds as it loses energy and increases in density. These gas clouds move along orbits within a proto-galaxy, and when clouds collide, gas experiences compression being compressed along a shock. The first stars of a galaxy are born directly of this process. A proto-galaxy evolves into a primitive galaxy with the emergence of the first rays of light from thermonuclear fusion. A group of young stars forming simultaneously remain embedded within their respective clouds of origin which, in turn, they begin to heat. At the end of the process, the heated gas is left to gradually disperse away, leaving behind a star cluster. The first stars to originate through this process in the Milky Way are still visible today in the globular clusters orbiting at large radii outside the galactic disk.

In galaxies where star formation continues, distinctive geometries begin to emerge; two of the more striking varieties are the spiral and elliptical. New stars generally still form in spiral galaxies but not in elliptical ones (Steinmetz). Both types start their lives with the same density enhancements present when the nascent universe first generated hydrogen atoms. But their respective initial distributions of stars originate at different speeds yielding appreciable differences in shape and later capacity for stellar generation. Our Milky Way galaxy is spiral where slow initial star formation did not consume the gas of a theoretically equivalent elliptical galaxy, giving it a spiral shape instead of the smooth, rounded appearance of an elliptical galaxy. The initial star burst of the Milky Way occurred 14 billion years ago. Gas that is abundantly seen even today falls into a spiral disk and drives star formation in areas of the galaxy where gas collisions occur.

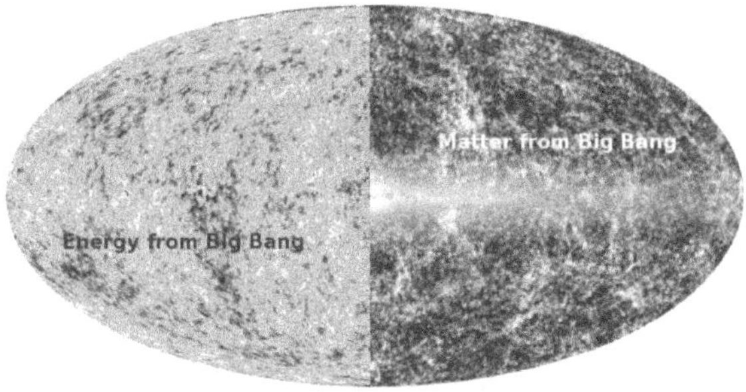

Figure 1 Then and Now. Energy and Matter from the Big bang. An image of the universe just 379,000 years old as captured by Wilkinson Microwave Anisotropy Probe (WMAP) project (Left half) overlaid on a map of galaxies (Right Half) in the 2MASS survey.

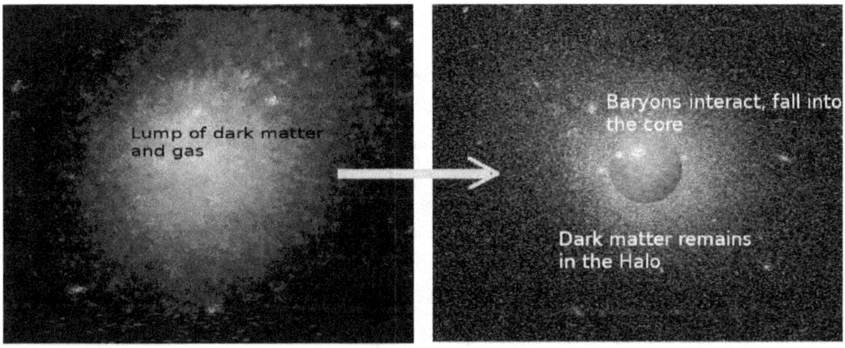

Figure 2 : Formation of a proto Galaxy from a lump of gas and dark matter (left) Gravitational collapse separates the lump into a core and a Halo (right)

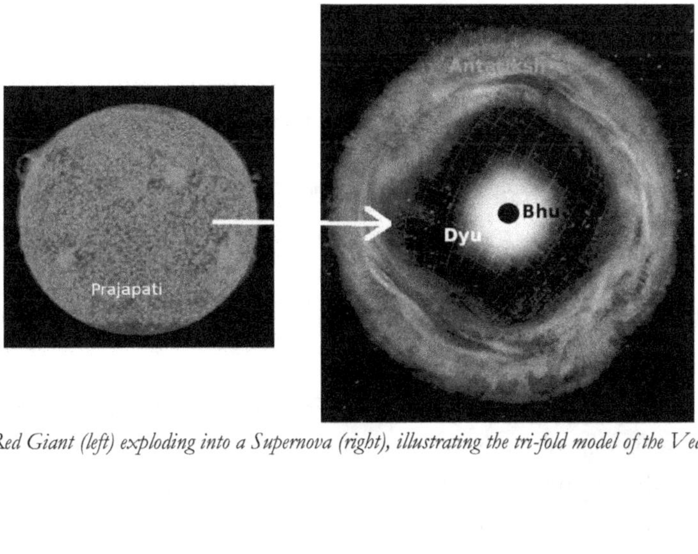

Figure 3 : Red Giant (left) exploding into a Supernova (right), illustrating the tri-fold model of the Vedas

The Active Universe:

The most massive stars end their lives in supernovae, explosive stellar destructions. A large star nears this phase of its life upon exhaustion of the fusionable elements in its core. As the outward radiation pressure of nuclear fusion gradually decreases, a star may collapse under its own weight. Falling hydrogen from a collapsing star's outer envelope collides with its core, igniting explosively. Having initiated the production of heavier elements, the fusion process gives way to the ejection of a shell of enriched gas into the galaxy's gas supply. Older galaxies are thus chemically evolved and have a richer mix of heavier elements. Life on earth depends on the presence of these elements; the fundamental composition of our human bodies derives directly from the supernovae of the past.

The primordial universe initiated its existence through a separation of the whole; what existed as oneness assumed disparate identities. At the very onset of the universe, the plasma preceding the initial separation contained the subtle element (photons) and the gross element (baryons and dark matter), galaxies forming from the latter. The triadic creation process described in the Vedas resonates with this characterization. According to the Vedas, "*Bhu*," the dense matter dominating the material existence was separated from "*Dyu*," the domain of subtle energies by "*Antariksh*" the intermediate space. Before creation, *Bhu* and *Dyu* exist as one. By the Vedic tradition, this triadic creation repeats at every level of existence.

Triadic stratification recurs in the process of the formation of a star. The densest material of a collapsing molecular disc coalesces into its associated planets, the *Bhu* aspect, while the subtlest material radiates out as light from the central star itself, the *Dyu* aspect. The atmosphere surrounding a planet is part of the interstellar space representing the *Antariksha*. Similarly, the Vedic *Bhu, Dyu,* and *Antariksha* aspects are found in the process of stellar destruction in the form of dense collapsed material, the surrounding accretion disk, and escaping radiation respectively.

Powerful energies move the universe from one steady state to another. The Vedic seers describe the transformation process in their unique style. In the Vedas, the primordial plasma is the Cosmic Egg – *Prajapati.* The initial oscillation that set creation into motion was cognized as an intention arising in the mind of *Prajapathi. Vayu*, one of the Devata forces, assisted in the ensuing physical separation, culminating in a *Bhu Antariksh Dyu* triad pictured in Figure

3. To the Vedic seers, the creation process is permanently ongoing; this triad configuration repeats in every phase, macroscopic to microscopic, of the universe. Our Sun is a notable instance of the *Dyu* element of the triad, the Earth its associated *Bhu*. Accordingly, the Sanskrit name for Earth is *Bhu-mi*.

The universe of perception

Birth is a movement from darkness to light. The nascent universe experienced the spread of cosmic background radiation in its beginnings, the birth of the first galaxies bringing light. On a smaller scale, the birth of a star floods its planets with light and energy with consciousness awakening to an experience of this light with the arrival of life forms on a planet. The Vedic seers attribute the movement from darkness to light to the *Adityas,* a class of *Devatas*. The Devatas *Vayu, Agni,* and *Indra*, among others, represent the principles, energies, and forces supporting the active Universe.

No single discovery in science has ever been the ultimate one. Research continues after every breakthrough, however brilliant. This basic idea applies to the works of the Vedic seers as well. Many scientific discoveries arise as sparks of intuition, shaped by the human intellect into scientific theory. Science deals fundamentally with the description of reality through models and theorized laws of nature; similarly the collective cognition of the Vedic seers shaped their understanding of the universe – one described by the Vedas as being driven by *Devata* impulses.

According to the Vedic tradition, the knowledge acquired even by its most advanced practitioners was limited. None claimed to have an absolute understanding of the universe, but rather, the tradition encouraged its students to build on the knowledge of their time, to stand on the shoulders of giants, as it were, and to develop new ideas through analysis, meditative experiences and dialogues with experts.

One branch of knowledge in particular benefited greatly from this open ended, constantly inquisitive vision the tradition espoused so strongly: Vedic astronomy.

2. Sorting Through Chronologies

Historians agree that the Nakshatra system belonging to Vedic astronomy has existed since antiquity while noting that many details may have been lost over the millennia. Its calendar system, the Panchang, has survived; in it, the Nakshatras are one of the five markers of time. The Panchang is consulted in determining the timing of various Hindu rituals.

The oldest[1] available Jyotish text expounds on a few calendric principles related to the timing of Vedic rituals,[2] using Nakshatras in its examples. Colonial scholars understood these references to indicate the age of the calendric system and the related Vedic rituals. To them, these references became not only proof of the age of the Jyotish branch but the Vedas themselves. This controversial approach dates the Vedic civilization many centuries after the Harappa civilization. This chapter introduces the reader to the genesis of this controversy.

Linguistic researchers of the eighteenth century segregated the Siddhantic text from the Vedanga Jyotish literature. Based on the writings of a Persian scholar from the eleventh century who accompanied the first Moghul invaders of India, they defined a period called the Siddhantic period, starting with the famous Indian astronomer Aryabhata. In doing so, they assigned Lagada's *Vedanga* text to an earlier age referred to as the Jyotish period. The start of the Vedanga Jyotish period was determined through the interpretation of a very limited sampling of references found in Lagada's text. Early linguists' limited attempts at defining chronology became widely accepted as accurate despite later disagreement from researchers studying the entire body of Vedic literature as opposed to a single document.

In order to understand the errors made in defining the chronology of Vedic astronomy, one requires some understanding of the Siddhantic class of text. The Vedic education system relied on their six classical limbs of learning: the six Vedangas, Jyotish ranking among these. All six were critical to the success of transmitting knowledge of the Vedas. The six Vedangas -- *Siksha, Chhandas, Vyakarana, Kalpa, Nirukta*, and *Jyotish* -- deal with increasingly complex subject matters related to Vedic learning.

---

[1] Lagada is the author of the oldest available Jyotish text

[2] Some of these Vedic rituals have fallen out of common use centuries ago

They are best understood as manuals of best practices, the means to master the material contained in the Vedas. The Vedangas themselves may well have been revised over the centuries to meet the needs of the era. The oral tradition through which Vedic knowledge is transmitted customarily retains the latest revisions. The contributions of earlier authors can thus be inferred only through their respective quotations.

The authors of the Siddhantic texts make ample reference to Lagada's ideas. While very few verses of Lagada survived directly into the current Rig and Yajur Veda texts, more may have been available to Aryabhata in the fifth century. Aryabhata both quoted Lagada's ideas and incorporated them in his own work. The Vedic tradition of an author of a new revision crediting an immediate predecessor is seen extensively in the works of Yaskacharya, the author of *Nirtukta*, another of the Vedangas, further discussed in Chapter 7. Thus, any realistic chronology of Vedic astronomy must understand the import of textual references found in the works of famous revisers of astronomical work.

Jyotish held an important place in Vedic civilization. If the Vedas were the human body, Jyotish was its eyes. As a study, it has proven indispensable in determining the timing of Vedic rituals, both contemporary and ancient. The ability to predict astronomical permutations requires abstractions both mathematical and scientific. The main body of the Vedas was sustained through oral tradition, by families dedicated to its preservation. However, furthering astronomical study required structures and instruments to observe the sky, the construction of which required royal patronage.

Errors could easily creep into the science of construction of geometrical structures during early times, before the advent of the standardized system of writing. The *Sulba Sutra* (Price) texts of the Vedanga called the *Kalpa*, describes the construction of Vedic altars with amazing intricacy. A few Sulba Sutra texts have survived intact into modern times because the geometry needed for *Kalpa* constructions is relatively simple. However, spherical geometry is an important aspect of astronomy and is difficult to teach even today without using physical models.

Distinction among Vedic Knowledge systems

Vedic knowledge is split into three branches: the *Shruti*, the *Smriti* and the *Puranas*. Shruti was given the highest priority among the three; it consists of the four Vedas. Each Veda has multiple *Shakas* (branches). A single family was able to preserve a single *Shaka* for generations, Shaka sized just to the right proportion such that a new generation of students could achieve mastery over it from the ages of eight to fifteen. While the modern education system relies on written and, increasingly, electronic material, the Vedic system was based heavily upon listening and memorization. It heightened the human brain's auditory, memorization, and vocalization capabilities. Student communities lived in partial isolation from the rest of the society while cultivating focus and sharpness. The society gradually structured itself around the familial profession of chanting and maintaining the Vedas. Multiple families preserved *Shakas*, and over a thousand *Shakas* were thus maintained for generations within each kingdom.

Students could pursue education beyond the scope of their familial *Shakas;* some also proceeded to learn the *Smriti* texts. Top priority, however, was assigned to *Shrutis* because they contained the cognized mantras and the knowledge of the physical realities of the cosmos, relevant at all times. Jyotish is fundamentally a part of the *Shruti* tradition. *Smriti*s contain guidelines relating to the optimal structure of society to support the preservation of cognized and acquired knowledge.

Variations in these guidelines, according to the terrain and geography, suggest that the *Smritis* were far more flexible when the Vedic civilization flourished in the Indian sub-continent. The early *Smritis* are attributed to prominent personalities in society such as royalty. Buddhist and Jain thought swept the sub-continent before the start of the Common Era. Many Vedic practices declined in frequency and scope during that time. Astronomical models, inherently necessary for the timing of these practices, thus became neglected for centuries until *Adi Shankaracharya* revived Indian society's interest in Vedic traditions. The analytical, logical approach of the Aranyaka[3] section of the Vedas gained popularity. Arts, artisans, and architecture flourished after society amalgamated Vedic traditions with the enshrining of statues which had gained in popularity during the Buddhist and Jain era.

The *Puranas* were revived as well and together with Smriti and Shruti complete the Vedic knowledge system. The Puranas convey knowledge through stories.

---

[3] The Upanishads use the dialogue style of presenting knowledge which is a hallmark of the Aranyaka section of the Vedas.

Stories and folklore in societies across the world attract the attention of the average citizen through relatable characters with superhuman achievements; the *Puranas* are no different. In it, characters frequently overcome life challenges through discipline, compassion, and valor. The heroes in the Puranas exhibit the ideals laid out in the Smritis. The Puranic tradition strengthened the moral fabric of society and reinforced people's faith in the Vedas. The stories in the Puranas take the listener's attention beyond the scope of their own existences, containing such large spans of time as the *Yugas, Manvantaras* and *Kalpas.* They weaved together stories over the eras, the fabric of a civilization spanning many kingdoms and many ethnic populations; they increased the psychological comfort of the common man and his trust in the system.

The Purana tradition survived through the Buddhist and Jain periods better than its Shruti counterpart. Knowledge of more complex Vedangas such as Kalpa, Jyotish, and Nirukta, became concentrated in limited centers of learning like the city of Ujjain, and the scholars could thus easily fail to grasp the original intent behind the division of the Shruti and Purana texts. As the centuries progress, one finds undercurrents of confusion in the works of astronomers. *Al Biruni*, the Persian scholar who visited India in the twelfth century, picked up on this conflict immediately, highlighting them in his *Tarik E'Hind.* European linguistic researches used the commentary of Al Biruni to brand the Siddhantic tradition as deviant from earlier works.

The Siddhantic approach to astronomy is an integral component of the Jyotish tradition. Aryabhata, among the first of the Siddhantic authors, states this explicitly in his works. Al Biruni wrote of the astronomer *Varaha Mihira*, a student[4] of Aryabhata, stating that Varaha Mihira knew of five Siddhantic schools. Linguistic scholars of the eighteenth century rely heavily on Al Biruni's conclusions about the origins of the *Romaka* and the *Paulisha Siddhants.*

A second myth relates to the observational skills of ancient Indian astronomers, originating from the fact that the Vedic experts available to Al Biruni were more likely to answer his questions from a Puranic angle, rather than a Jyotish one. Without a clear distinction between the two, Al Biruni drew erroneous conclusions about the level of competenacy of local experts. Al

[4] There was also a Varaha Mihira who lived a few centuries earlier but the common understanding is that Al Biruni was writing about Aryabhata's student.

Biruni expressed his frustration in acquiring explanations from local experts on the correspondences between Nakshatras and stars in the sky.

The Islamic world had been the conduit of information flow from India to the west for many centuries. The twelfth century works of the Persian author *Al Qazvini* provide examples of the clues that the colonial world stumbled upon but chose to ignore.

The eighteenth century and the Asiatic Society

The British established the Asiatic Society of Bengal in Calcutta to facilitate translation of Vedic literature. Its associated scholars possessed an understanding of the uniqueness of these Sanskrit texts, many of them adherents to the *diffusionism* doctrine proposed by Isaac Newton (Schaffer, 1980). Diffusionism, in essence, proposed that significant knowledge held by ancient civilizations had been lost during the plunder and upheaval of the Middle Ages. William Jones, Reuben Burrows, and William Hunter, among other members of the Asiatic society, were intent on leading local scholars from using the Arabic language of the status quo -- then the official language of India during Moghul rule -- and on calibrating local understanding of ancient knowledge against its European counterparts.

Reuben Burrows hired Tafazzul Hussien Khan, a talented Shia Muslim linguist from Persia hailing from Lucknow, India, to translate Newton's Principia. Tafazzul became the Asiatic Society's intermediary in the procurement of copies of Sanskrit literature. From at least the twelfth century onward, Persian scholars were well-recognized for their skills in translating works between Sanskrit, Persian, and Arabic, a trend started by Al Biruni. Tafazzul and others often served as principal translators, having developed the linguistic skill necessary to work directly with local Pundits.

Just a few years into the project, William Jones, the Asiatic Society of Bengal's founding president, published an influential paper aligning the history of the Hindus with popularly accepted Christian chronological tables. The paper had an immediate influence on local scholars; Jones' writings indicated that ancient Sanskrit records could be better understood through the context of European scientific achievements. Samuel Davis, Jones's assistant on the subject of astronomy at the Society, demonstrated how the ideas of Surya Siddhanta complemented the Greenwich nautical almanac.

Jones' translations were well received by local scholars. Newtonian thought had flourished in Bengal during the formative years of the Asiatic society, and British-educated intellectuals of India could, for the first time, better understand Vedic literature through these translations. Newton himself had once claimed that the theory of heliocentrism was once known to elite groups in some regions of the world and was lost in Europe during the Middle Ages. Understandably then, the age saw an increase in native sense of pride in traditional achievements.

However, the chronology of the development of science in India was still forced to fit the prevalent scriptural views of the seventeenth century Europe. As flawed timelines quickly gained orthodox status, Colebrook (Colebrooke, 1817), Playfair (Ray, 2009), and other contemporary indologists proposed more appropriate methods. But dissenters received severe scholarly rebuke, with volumes of subsequent research and translations unable to make a dent in the restrictive chronology scholars like Max Muller wanted established. The views of Colebrook and others were slowly pushed to the wayside.

An unfortunately illustrative case is that of Thomas Pearse, the colonel in the Artillery division of the East India Company. Through his travels in British India, he encountered multiple baffling astronomical achievements of ancient Indian scholars. In a correspondence with Joseph Banks, Pearse included an ancient illustration of Saturn with six hands and a seventh hand holding a ring. We know today that Saturn has six moons larger than 300 kilometers, most famously seen in the iconic composite photograph taken by *Voyager 2* in its 1980 flyby of the ringed planet. Pearse expressed his view that the natives had observational skills that far outstripped contemporary assumptions.

William Jones ridiculed Pearson's ideas while delivering his obituary.

Events in the Eleventh Century

Islam arrived in India, by way of Persia, with the invasion of Gazni of Mahmud. The locals, especially the Pundits, were wary of sharing ideas with brutal invaders who showed very little tolerance to the native brand of spirituality. The scholar Al Biruni accompanied Mahmud of Gazni, in many ways a breath of fresh air compared to his compatriots. Because of his familiarity with Indic traditions, then practiced in Afghanistan, Al Biruni was successful in establishing contact with local Pundits and began translating works from Sanskrit. In Persian he composed his famous *Kitab-ul-Hind*, a detailed study of

Indian customs, traditions and way of life. The *Kitab-ul-Hina* grew popular with linguistic researchers who used it as a primary source of information on the state of astronomy in India.

Al Biruni did not trust prior translations of works from ancient India, instead preferring to validate his understanding against first hand documents. Sensing his openness, local Pundits were willing to assist him. Al Biruni gradually worked through the analysis and theories of numerous Indian astronomers, particularly the works of Varaha Mihira and Brahma Gupta from the sixth century.

From his work during this period, linguists conclude that five Siddhantic schools existed during the time of Varaha Mihira. But while Al Biruni attributes one of the Siddhantas to a Greek scholar and another to a Byzantine author, the texts of these two Siddhantas clearly name local astronomers as their original authors. Colonial scholars popularized Al Biruni's opinion and attributed any conflicting proof in the texts themselves to conspiracies by contemporary Pundits.

Al Biruni highlighted Brahma Gupta's critique of a work of Varaha Mihira, the non-astronomical *Brihat Jataka*, espousing the supremacy of the Siddhantas over Puranas. Varaha Mihira likely made this claim with a view to restoring balance among the three types of Vedic knowledge, with many of his contemporary astronomers more prone to favor Puranic ideas over those of Jyotish. Colonial historians ignored this course of events, reverting to popular stereotypes of Hindu cosmology and claiming the more allegorical, metaphorical nature of the Puranas -- turtles supporting the weight of the universe -- as evidence of primitive thought while ignoring the striking astronomical advances of Jyotish methodology (Minkowski).

The lineage of astronomers beginning with Aryabhata and ending with Bhaskaracharya (Bhaskara II) in the twelfth century are recognized today for their mathematical brilliance. Moghul rule rapidly expanded through India, and Arabic studies superseded their Sanskrit counterparts in royal courts. The study of the Vedas in Northern India was a direct casualty, with an increasing number of astronomical works during the period focused on mere restatement earlier discoveries.

The pervasive influence of Arabic culture reached even the Panchangs, which began to include Hijri calendric information. Raja Jai Singh, the king of Jaipur and an avid astronomer, eventually undertook the challenge of reviving observational astronomy in the early eighteenth century. His effort in building a modern observatory, combining the best of western astronomy with Siddhantic guidelines, is well documented. Singh's observatories stand today as the only physical evidence of the observational capabilities of ancient astronomers.

The fifth century and Aryabhata

Aryabhata, the fifth century astronomer, was a contemporary of Varaha Mihira whose works find prominent mention in Al Biruni's Tarik E'Hind. Aryabhata revived ancient Siddhanta concepts, especially from the Surya Siddhanta. His work is largely mathematically based, attuned to observations of the sky. Aryabhata condensed his ideas into a set of 108 verses, intending to aid memorization, in his famous work called *Aryabhatiyam*. Astronomers of the following centuries wrote ample commentaries on his works, with Aryabhata also establishing the tradition of revising planetary tables through the centuries. He revived interest in a new generation of astronomers to review all schools of Siddhanta. The most notable was the astronomer Brahmagupta, whose work focuses on the Brahma Siddhanta.

Aryabhata simplified astronomical formulae by introducing the concept of counting total planetary rotations from a fixed conventional point in time. He achieved this by examining planetary positions at the start of the Kali era, BCE 3102 and close to the zero point of Vedic astronomy, in the *Revati* Nakshatra to create constants for his formulae. The Surya Siddhanta mentions that computational errors creep in overtime and that the constants assumed to represent rotation values needed periodic revisions. Aryabhata established the importance of periodic revisiting of these constants based on new observations, but this practice may have waned in the centuries preceding his lifetime.

Aryabhata's use of mathematical models in astronomical predictions may have catalyzed the growth of mathematics as a field of science in India. Through the commentaries of his disciples Bhaskara and Brahmagupta, some details of his specific observational instruments are available, in addition to his own *Arya Siddhanta*. Still, much knowledge was lost in the subsequent centuries including the construction and use of his instruments and the specifics of his mathematical tools. We know, for instance, that Aryabhata used the decimal system only through the knowledge that numbers were alternatively represented through the Sanskrit alphabet (Wilke, 2011).

BCE 3102 and the start of a cultural experiment

According to Aryabhata, the planets were in alignment on the 17th of February, BCE 3102; this is the date of the start of the Kali Era. Important historical events in the Puranas are frequently supported with astronomical signatures. The description of the beginning of the Kali era, for example, is supplemented with data on the occurrence of triple eclipses in the preceding decades as well as the positions of the planets in their respective Nakshatras.

Modern astronomical software helps identify specific times when the particular pattern of triple eclipses referenced in Puranic texts could have occurred, with modern research delving into the guidelines behind the timing of Vedic rituals The researcher K.D. Abhyankar points specifically to the striking visual organization of the planets in the night sky towards the dawn of January 10, 3104 BCE (Abhyankar K. , 1996).

The event was visible over a wide swath of the sub-continent and likely left a collective impression, with the entirety of Vedic civilization both preparing for a winter solstice ritual two days later and concurrently observing the night of *Shiva Ratri*, an important dusk to dawn celebration which continues to modern times.

Clay seals excavated from the ruins of the Indus Valley civilization may be representative of the solstice ritual. Two years later, the morning of February 17th BCE 3102 dawned in the Yoga Tara star of *Revati* Nakshatra, all the planets contained within a 40 degree span; this then became the zero point of Vedic astronomy.

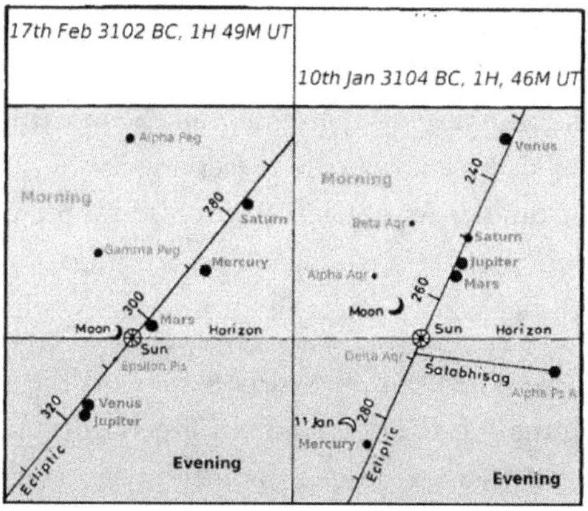

Figure 4 Position of Sun, Moon and Planets in Feb 3102 BCE and Jan 3104 BCE

As early as the 1780s, the Western researchers Playfair (Ray, 2009), Bailly, and Jacobi, made note of these nuanced astronomical references in Vedic literature. They accordingly proposed a far deeper, more advanced chronology for the Vedic civilization. According to Playfair, Vedic astronomers had an understanding of the equation of the sun's center and the obliquity of the ecliptic, and moreover, evidence pointed to the civilization's knowledge of precession, a phenomenon impossible to recognize without multiple centuries of observational data. Subsequently, Playfair supported the dating of the Vedic civilization to 4300 BCE. With local scholars espousing the idea that the Vedas were originally compiled around the start of the Kali Era, only the elongated, older chronology of Playfair could explain both this timeline of compilation as well as human observation of the astronomical signatures recorded in Vedic texts.

The later claim by colonial scholars of fabrications and after-the-fact insertion of advanced astronomical data into Vedic texts by local Pundits seems worth pondering until one realizes that such assertions ironically credit would-be conspirators with knowledge of Newton's and Kepler's laws long before they were discovered! The idea that local Pandits of the 17th century could successfully simulate and recreate the position of stars and planets in the night sky multiple millennia into the past -- a feat only achieved by modern astronomy following the advent of the computer -- is indeed farcical.

The huge volume of Sanskrit literature already present in colonial times could not realistically have evolved under the post-Indus valley chronology assigned by early Europeans to the Vedic civilization. That an extensive cultural endeavor aimed at the preservation of knowledge was initiated five millennia ago is likely closer to the truth. With the Shruti alone containing 1180 Shakas, a new student took eight years to master a single Shaka. In summary, students invested upwards of six billion hours in preserving the Shruti*s* from the advent of the system to the time of Aryabhata. Written script began to move towards standardization around his era, though memorization was still emphasized because he chose the succinct *Sutra* style for his main work.

The enterprise of memorization has carried forward the Shruti, Smriti, and Puranas along with the subsidiary literature for centuries. In that sense, Vedic civilization offers historians a unique opportunity to study human societies from the perspective of memorized, oral tradition entirely sustaining culture, customs, and knowledge base.

Indian history books describe the ancient universities of *Taxilla, Nalanda, Pataliputra*, and *Ujjain;* when he began to compose his works on astronomy in his twenties, Aryabhata moved to the University of *Kusumapura*. Such institutes of higher learning flourished well into the twelfth century, documented by the Chinese scholar Fu Xian on his visit to India. Volumes of material in the libraries of these universities were lost to fires during the first waves of the Moghul invasion of India. Modern day anthropologists have pushed the evolution of intelligent humans worldwide backward in time by several millennia, yet the accepted history of the Vedic civilization has not inched beyond the threshold set by eighteenth century historians. An arbitrary, inconsistent chronology has sadly continued to divert scholarly attention from the scientific content of the Vedas.

3. Confirming historic Evidences

The Yoga Tara (Lancelot Wilkinson, 1891) aspect of Nakshatra astronomy is mired in mystery with many texts describing the construction of astronomical instruments unpreserved or lost entirely The system of *Yoga Tara* helped astronomers revise calendars and verify the accuracy of their models with observations of the sky.

Independent India, as part of its first governmental initiatives, formed a calendar reform committee in the 1950s, identifying replacement Yoga Tara stars for a handful of Nakshatras. The formation of this committee underscores the importance of the Yoga Tara in the formulation of the Vedic calendar. The last existing documentation of Yoga Tara and the observational aspect of Nakshatra astronomy is found in the works of *Samanta* Chandrashekar, the naked eye astronomer. This chapter introduces his work, as well as the much heralded project of Raja Jai Singh seeking to revive the lost tradition of astronomical observatories. Singh hired a diverse, talented pool of Jesuit, Islamic, and Vedic scholars to create instruments illustrative of instrumentation principles elucidated in *Siddhantic* literature. The chapter ends with a short discussion on the evolution of the Vedic calendar system, introducing references to Nakshatras in the Vedanga Jyotish text.

Many associate the Yoga Taras more commonly with astrology, but their significance related also to the astronomical observation of the position of planets within a Nakshatra arc on the ecliptic. Yoga Taras create a framework for the sky that facilitates identification of the boundaries of a Nakshatra along the ecliptic, along the celestial equator, and at the horizons.

Critics have historically raised three objections to the practicality of the Yoga Tara system: a) Some stars of the set are too faint, (b) Most stars do not fall along the ecliptic, and (c) Three stars fall in territory closely adjacent to neighboring Nakshatras. The works of Raja Jai Singh and Samanta Chandrashekar are thus imperative in addressing these concerns, and additional research is required to fully understand and reconstruct these instruments.

Working models

Ancient astronomers of India used the Yoga Tara group of stars to make observational corrections to their calendar models. A tradition of periodic calendar corrections appears to have been sustained since the time of Aryabhata. Through statistical analysis, the mathematician Roger Billard illustrates that these periodic corrections have led to improved accuracy upon each iteration. Without observational assistance, these revisions for accuracy would of course be rendered impossible.

Opposed to Billard, the linguist David Pingree supports the orthodox opinion held for the past two centuries that ancient Indian astronomers lacked observational skills, citing ambiguity as to observational instrument construction. Van der Waerden analyzes the disagreement between the two sides in a paper in 1980 (van der Waerden, 1980).

The work of the twentieth century astronomer Samanta Chandrashekar stands as further proof that the Yoga Tara system and associated instruments were successfully used by ancient astronomers (Naik). Chandrashekar built simple approximate instruments and published models of reasonable accuracy in 1899. Using the Siddhantic texts as his guide and borrowing ideas from Sanskrit works of earlier centuries, he worked largely bereft of access to newer techniques available to his contemporary European and Middle Eastern astronomers.

A summary of the general nature of Chandrashekar's instruments and their constructions is presented here, extracted from his works (Upadhyaya, 1998-08). Through careful observation with his self-constructed instrumentation, Chandrashekar achieved startling accuracy in his predictions. Subsequently, he has been acclaimed as the greatest naked eye astronomer of modern times. His instruments highlight the distinct possibility that similar, if not far more advanced, observations were possible in earlier millennia.

Chandrashekar's instruments fall into two categories. The first is a spherical device, assisting the tracking of planets, using fixed stars as a guide. The second helps track the passage of time.

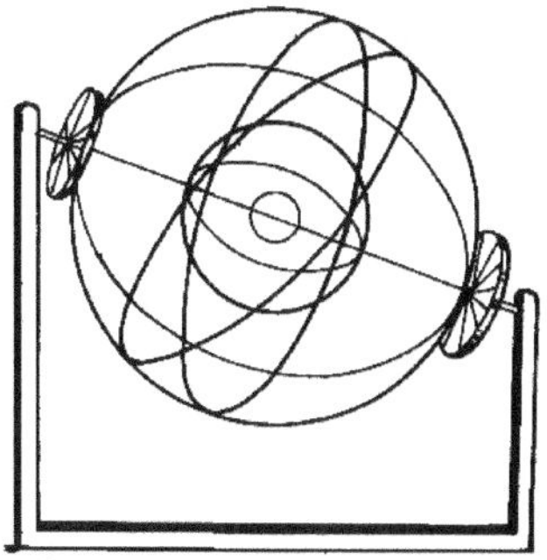

Figure 5 One of Samantha Chandrashekar's sky observation instruments

The works of astronomers dating back to Aryabhata contain references to various spherical instruments, but details of their construction lie hidden in the peculiarity of the style of their presentation. The innocuousness of the traditional naming convention -- for instance, the "man" instrument or the "peacock" instrument -- belies their respective purposes. Chandrashekar's research of Sanskrit documents and application of astronomical genius were instrumental in his reconstruction. Like various authors before him, Chandrashekar alludes to lost versions of instruments capable of tracking stellar motion with further automation. His own models required an additional, separate time tracking device that allowed him to determine when the instrumentation sphere, depicted above from his book Siddhanta Darpan, required repositioning. The instrument itself is shaped like an armillary sphere similar to those used by western astronomers before the advent of telescopes. Using his tools, Chandrashekar revised coordinates for the Yoga Tara stars. He invented new predictive models for the movements of planets, and Panchang calendars based on his models continue to enjoy popularity in his home state of Orissa in India.

While Chandrashekar followed the basic structure of Siddhantic models in his use of planetary revolutions from a fixed starting point, he deviated from Aryabhata by setting the fixed point to the start of a Mahayuga cycle instead of the start of Kaliyuga. This, according to him, increased accuracy in his calculations. He may have been the last astronomer to have used the Yoga Tara system (Abhyankar K. , 1991) for observational corrections. Telescopic observations in modern times are dependable and accurate, with the boundaries of Nakshatras easily located via modern telescopes. The importance of Yoga Taras has thus been relegated to the field of astrology.

The Stone Observatory

Raja Jai Singh (Virendra Nath Sharma) sought physical implementation of observational instruments specified in Siddhantic texts, and employed Indian, Jesuit and Moghul experts to iterate upon Siddhantic methods to reconstruct functional segments of ancient tools. The massive stone masonry in the observatories of New Delhi and Jaipur in India are proof of his expertise. In Raja Jai Singh's instruments, we see instrumentation concepts described by authors spanning back to Aryabhata. The components of Singh's observatory, though of a larger scale, permit us to envisage simpler instruments functioning at equivalent levels.

Chakra Yantra: Used to identify Meridian Pass Time and the declination of planets. It contains two circular structures framed on stone pillars and has gradations to measure declination. The circular structures can revolve parallel to Earth's axis and point towards the pole. The circular structure has a hole in the middle in which another instrument can be placed to make observations. One of the instruments that can be mounted in the center hole is a narrow metallic tube whose extremes touch the gradations on the circular structure. A celestial object can be viewed through the narrow hole while its position is noted. The southern disc indicates the meridian pass time of a celestial object.

Kranti Vrita Yantra: Used for direct measurement of celestial latitude and longitude. It consists of two brass circles, one capable of rotating in the plane of the celestial equator and the other in the plane of the ecliptic.

Kapali Yantra: Consists of a pair of bowl shape hollows in the ground. Each has gradations on its horizon for measurement. A ring is positioned with wires in the middle of the central hollow. The shadow of the ring provides Azimuth, Altitude, Meridian Pass Time, Declination of sun, and the local time. Lines drawn on the spherical surface of the hollow cavity and its gradations are used for taking measurements. One of the cavities is utilized for solving astronomical problems graphically. Jai Prakash Yantra is an innovation of this instrument. It demonstrates the "doctrine of sphere," illustrating the apparent motion of the Sun. It is a great educational tool for learning astronomy.

Rashi Valaya Yantra: A group of twelve instruments, with gradated quadrants on both sides. It provides direct determination of celestial latitude and longitude. They resemble, in principle, another instrument called the Samrat whose quadrants represent the celestial equator. Unlike the Samrat, here a quadrant represents the chords of the ecliptic. The use of a chord and the principle of lining up against a sky object using a thread is found in the descriptions of the Surya Siddhanta. A chord can easily represent a Nakshatra segment instead of a Raashi segment.

Dakshinobhitti Yantra: This instrument is mainly used for observing the altitude of objects in the sky. It is a big wall, placed in the north to south line of the meridian. The instrument is inscribed with two quadrant arcs on its east and west faces. Steps reach close to the gradations. At the center of the quadrants and arcs, small pegs are provided to tie a thread whose other end can be aligned with the object to be observed. The thread passes over the gradations on the east-west faces. The use of the chord and the thread is a principle seen in Siddhantic texts.

Table *1 : Pictures of instruments in Raja Jai Singh's observatory: Chakra, Kranti, Kapali, Dakshinobhitti, Rashi Valaya*

Chakra Yantra	Kranti Yantra
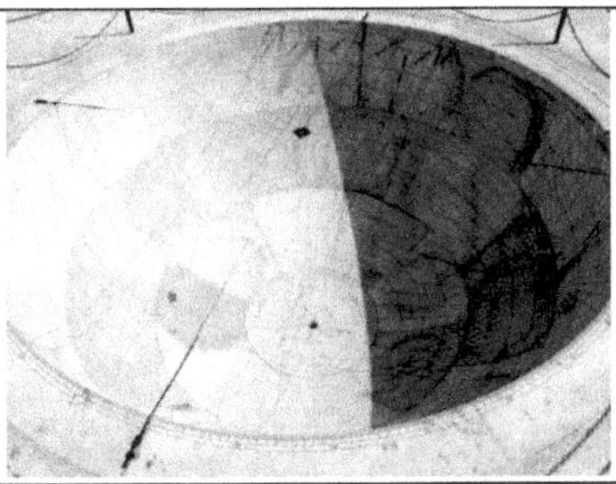	
Kapli Yantra	Dakshinobhitti Yantra

Rashi Valaya Yantra	

Dating of the Vedic Calendar

Aryabhata may have been the first astronomer to revive the tradition of updating the calendars popular during Lagada Muni's times, following a gap of several centuries. Lagada's works on Vedang Jyotish contain calendric prescriptions from the Rig Veda and the Yajur Veda. Aryabhata used 43 verses of Yajur Jyotish and 36 verses of Rig Jyotish in his model. The Yajur Jyotish calendar uses a 5 year Yuga cycle (Achar, 1997) to adjust the number of synodic months in a lunar year to align it with a solar year.

Early Indian researchers, in attempting to reconcile the five year cycle described in the Yajur Veda with a similar cycle described in the Rig Veda, made fourteen changes to Rig Jyotish verses to achieve "meaningful" interpretation. The changes, they claimed, were required because the text had become corrupted.

Unsatisfied with this explanation, PV Holay pursued a more detailed study (Holay) of the differences between the verses of the Rig and Yajur Jyotish. In a paper on the subject, he explains that Rig verses encode the Nakshatras positions of the solstice moons for 19 years in an arithmetic series. The Nakshatra *Shravishta* is the first in Lagada's illustration of the principle detailed by Holay. The following 19 year cycle begins with the preceding Nakshatra rather than Shravishta, and therefore, the 19 year cycle is called a Yuga period. Other scholars have taken the arithmetic series to indicate a Yuga cycle of five years. These verses illustrate the intricately coded style of the Vedas. One particular section encodes the concept of the 19 year cycle using 38 Sanskrit letters, borrowing the first letters from the name of the Nakshatra. Without an understanding of the code, the meaning is lost entirely.

The verse that references the Nakshatra Shravishta has been used by scholars to date the age of Lagada as well as the Vedas. When Lagada first wrote the verse, the winter solstice was presumably in the *Shravishta* Nakshatra, a two millennia long period. Narrowing this time frame necessitates an additional astronomical marker: another of Lagada's verses that encodes the name of the month in which the solstice took place. Two alternate interpretations of this code give answers separated by multiple centuries; they derive from alternate conventions of beginning the lunar month on either the full or new moon. Holay's interpretation uses the full moon convention and indicates that the winter solstice occurred in the *Magha* month. His dating of Lagada's time allows a thousand more years for the development of subsidiary Sanskrit literature before the time of Buddha.

Concepts related to the predictability of the solstices were built into complementary Vedic ideas -- a solar year consisting of two solstices, a sidereal year, a Savana year consisting of twelve months of thirty days each, and a lunar year consisting of twelve lunar months. BG Sidharth (Sidharth, 1998), another researcher of the Vedic calendar system, explains the astronomical principles behind the codes of complementary concepts. The number of Devas in the Rig Veda is listed as 11, 33, 303, or 3003 in various verses. The sequence 33, 303, 3003 (moving toward 30,000,003) refines the approximations of the length of a sidereal year.

A famous Rik – a name for mantras in the Rig Veda -- of the rishi *Vishwamitra* in the third mandala of the Rig Veda notes that "3339 devas are going around Agni." BG Sidharth explains the figure 3339 as the number of intercalary[5] days in 297 years (33 X 9). The mean length of a solar year calculated over 297 years is 365.2424 days, an amazingly accurate result. The difference between a mean sidereal month and a synodic month calculated over that period is 2.20893. Holay suggests that through the 3339 code, Vishwamitra understood the calendric concepts of sidereal and lunar units.

The Vedas associate *Gandharvas*, a class of invisible energy beings, with all atmospheric phenomena as well as the Moon. The number of *Gandharvas* is noted as both 27 as well as 6333. BG Sidharth provides an elegant explanation of how these numbers encode the ratio of the length of a synodic month to a sidereal month. Lagada, in his Vedanga Jyotish, translates these codes into simpler calendric concepts while Siddhantic literature simplifies these concepts into mathematical formulae.

Lagada's work condensed complementary models of the Rig and Yajur Vedas, illustrating them with examples to help students understand the principles behind his predictions. His verses on the Nakshatra Shravishta enabled his students to predict new Yuga cycles based on the current configuration of the sky. Lagada does not mention the Yoga Taras presented by the Surya Siddhanta as an observational support system. But student astronomers clearly required techniques and instruments to recognize the Nakshatra boundaries integral to their observations and to use Lagada's formulae. Such knowledge may either have been widely known during Lagada's time or have been available in ancillary works similar to the Surya Siddhanta (Gangooly, 1989). It is also possible that parts of Lagada's works did not survive into modern times.

---

[5] A day or a month inserted in a calendar to align it with a solar year

Proof of earlier observational tradition is found in Buddhist and Jain astronomical treatises, which provide detailed instructions on the use of observational techniques and instruments. The simplest of these techniques relates to the use of 'arc chord geometry' on the ground. Some of the temples of the famous Ajanta caves align to the sun's position on specific days of the year, an affirmation of the accuracy of these techniques. Buddhist and Jain texts also show that architects had mastered planar geometry although knowledge of spherical geometry may have declined as practice of Vedic rituals tied to astronomical phenomenon such as equinoxes had by then began to wane.

R*ituals as a Calendar system*

Even today, most Hindu festivals are based on a lunar calendar. A lunar year falls short of its solar equivalent by eleven days. Many discontinued Vedic fire rituals were based on lunar and Savana years and not the solar year, with various calendar systems providing alignment between systems. The modern day Panchang, a widely used calendar system in India, adds an intercalary lunar month every four to five years to synchronize calendars.

Annual Vedic rituals were performed over a period of twelve months defining the boundary of a ritualistic year. The most common of these is the *Samvatsara Satra* which began on the winter solstice. An *Atiratra*[6] ritual filled the gap at the end of a Samvatsara Satra until the next. Thus, the Atiratra aligned the Savana year with the solar year. A variation of the Atiratra was used when the annual ritual was conducted over twelve lunar months instead of over twelve Savana months (30 days long). A *Dakshinayina*[7] ritual spanned thirty years; the back to back performance of Dakshinayina adhered to strict guidelines known as the *Agni Chayana Vidhi*[8]. These guidelines created a 95 year long ritual cycle.

The Atiratras adjusted a ritual year to the Solstices. Some historians therefore view them to be early attempts at a calendric system, believing shorter variations of the Atiratras to have evolved subsequently. This idea fits well with

[6] A 5 day ritual that took a Savana year from 360 days to the start of the next Savana year. This could be split into two shorter ones called *Ratantara Saman* and *Brihad Saman.* The *Atiratra* was 12 days when the 360 Savana days became replaced with 360 lunar thithis.

[7] *Dakshinayina* sacrifice dropped one lunar month in two parts after 15 years.

[8] *Dakshinayina* needed adjustment after three cycles. An agni chayana rule was introduced involving a 95 year cycle

the conventional historic view that ancient society did not have an elaborate calendar until the time of Lagada but also ignores the fact that the Pundits required calendars to conduct regular fire rituals celebrating life transition events. Such rituals were performed within families without community participation. It is therefore more likely that the longer fire rituals fell out of use due to the demand for more resources and coordination. The 30 year and the 95 year rituals could not be conducted without royal support. The shorter fire rituals simply adapted more easily to a transitioning society.

A team of researchers studying the details of the Atiratra fire altar (Price) concluded that detailed knowledge of geometry was necessary in the construction of the ritual's structures and assembly. The Atiratra fire altar is composed of alternating layers of bricks with the specific geometry shown in the figure below. Family fire altars did not require such complicated geometry. This too supports the alternate historical view that fire rituals that could not adapt to the changing times were discontinued. In line with this view, astronomical knowledge also diminished with changing customs

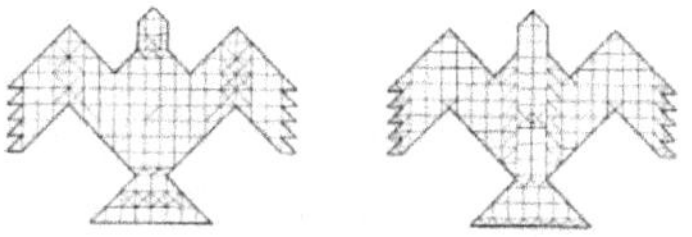

Figure 6 : Atiratra Altar - brick construction requires knowledge of complex geometry

The Stars of the Yoga Tara set

The Rishi *Agastya* is a colorful personality in Vedic literature. Tales from his life abound both in Sanskrit language works as well as in Tamil Sangam literature. Important stanzas of the Rig Veda, Mandala 1 are attributed to Rishi Agastya. An important annual Vedic ritual is associated with the sighting of the star Agastya, and early astronomers may have helped predict the timing of this ancient Vedic ritual. Agastya corresponds to the star Canopus, bright and visible only from the southern latitudes of India in ancient times. Over the millennia, it became visible from more northern latitudes as well. Changes to its latitudinal visibility make Canopus an important astro-archeological time marker.

The pair of Yoga Taras, *Apam Vatsa* and *Apah* longitudinally line up with the modern star Spica which is also the Yoga Tara of *Chitra* Nakshatra. Sidereal poles and the poles of the Jyotish coordinate system, fall on the meridian circle that links these three Yoga Tara stars. The zero degree longitude of the Vedic sidereal coordinate system is also part of this circle, with the zero point represented by the Yoga Tara of *Revati* Nakshatra. The heliacal rising of the bright star Spica coincides with the heliacal setting of the Yoga Tara of Revati. Heliacal observations were the most efficient method of locating the position of the sun along the ecliptic in ancient times. The choice of a faint Yoga Tara for Nakshatra Revati is justified by its 180 degree position away from the Yoga Tara of Chitra.

The Yoga Tara named *Brahma Hridaya* is the bright star Capella. Capella does not heliacally set in the Indian latitudes. It can be easily spotted in the night sky from anywhere in the Indian subcontinent due to its brightness. Brahma Hridaya lines up with another bright star in the same longitude, closer to the ecliptic. This second Yoga Tara star is called *Hutabuk*, another name for Agni, or fire. The third Yoga Tara named Prajapati lies just 5 degrees east of Brahma Hridaya and also does not set in the Indian latitudes. Tracking the rise of the star Hutabuk facilitates the observation of the neighboring Yoga Taras, which belong to the Nakshatras *Rohini, Mrigashira* and *Ardra.* An instrument such as the *Rashi Valaya Yantra* (see section on Raja Jai Singh) illustrates the principle of segmenting the sky for simplifying observations. Such an instrument could have zoned in on the Yoga Taras of Brahma Hridaya, Prajapati and Hutabuk to identify the rising and the setting of the boundaries of these Nakshatras.

The Yoga Taras of *Abhijit* Nakshatra (Alpha Lyrae), *Shravana* Nakshatra, *Shravishta* Nakshatra and *Uttara Badrapada* Nakshatra do not set heliacally in the Indian latitudes. The Yoga Tara of *Swati* Nakshatra (Arcturus) has a high northern latitude and lies longitudinally across from the Yoga Tara of *Bharani.* The principle of simultaneity of heliacal rising and helical setting of the Yoga Taras of Swati and Bharani applies to this pair as well. A spherical instrument could easily identify the Yoga Tara of Bharani despite its dimness, in a manner similar to the Yoga Tara pairs of the *Revati* and *Chitra* Nakshatras.

The Yoga Taras of *Pushya, Magha* and *Revati* fall on the ecliptic while the Yoga Taras of the Nakshatras *Krittika, Rohini, Chitra, Vishaka, Anuradha, Jyeshta, Mula, Purva Ashada, Uttara Ashada* and *Shatabhishaj* fall within a 5 degree from the ecliptic. Of these, the Yoga Taras of Chitra and Jyeshta have the best heliacal setting strength, quoted as 13 degrees of time[9]. The next group of Yoga Tara stars which belong to the Nakshatras Rohini and Vishaka have 14 degrees of time strength. The remaining four have a 15 degree of time strength. The Yoga Tara system is similar to the Egyptian Decan system though the Decan system is comprised of stars in 10 degree segments along the ecliptic. Yoga Tara has few stars that lie away from the ecliptic, with the remainder clumped into groups rather than spaced evenly.

The Yoga Tara named *Mrigavyadha* is also called *Lubdhaka*, the subject of a story in the Rig Veda. Lubdhaka is the star Sirius, and is the second brightest among the Yoga Tara set. Because of the star's large proper motion, it appears immune to the effect of precession shift. The Egyptian calendar system of 3000 BCE utilized knowledge of the proper motion of the heliacal rising of the star to set the first day of the calendar year. Similarly, the star served among the key anchors of the sky in the calibration of Jyotish instruments, though its coordinates have deviated considerably from the value noted in the Surya Siddhanta.

---

[9] An ancient way of measuring the brightness of a star based on how close to sunrise a star becomes invisible

4. Exploring Cyclical Time

Human beings have relied on cyclical changes in the surroundings to recognize the passage of time. The changing configurations in the sky provided the earliest methods for measuring time. Later innovations, tried to mirror changes in the sky with innovations on the ground. These innovations range from megalithic stone structures to miniature sun dials. The Nakshatra division of the sky provided a natural backdrop in the sky which the Vedic astronomers utilized to keep time in different units. The simplest unit of time is a day which on the modern calendar is counted between two consecutive midnights. A sidereal day on the other hand is counted between two consecutive sun rises. A sidereal day is the day unit in the Vedic calendar. A sidereal year consists of 366.2563795[10] of such day units (See important footnote below). A lunar month is an observable unit of month being counted between two consecutive new moons. Vedic astronomers split the lunar month consisting of 29.5 sidereal days into 30 equal units called *Thithis*. The Full moon shifts to a different Nakshatra in each month and this Nakshatra supplies the name for the lunar month. The solar year, measured from consecutive vernal equinoxes was also split into twelve month units called the solar months.

The Vedic astronomers brought the otherwise lunar centric calendar in alignment with the solar year by adjustments to the lunar month, a few accounts of which can be seen in chapter 3. Due to the lunar and the solar months being stepped side by side, scholars call the Vedic calendar system a luni-solar calendar. One notable feature of this calendar system is that the moon (the closest luminary) and the sun (the brightest luminary) arrive against the same backdrop of stars on one's birthday each year. It is thus a more genuine repetition of a celestial configuration on one's birthday contrasting to the mathematically terse configuration of earth returning to one's birthday position on its orbit around the Sun. The design for birthdays reflects the Vedic sentiment of the interdependence of human and cosmic intelligence.

The numbers six, sixty and three sixty are found to repeat in the Vedic units of time. A Vedic day is divided into sixty smaller units in comparison to the

---

[10] Vedic astronomy uses the diurnal rotation of the Earth to denote the length of a sidereal year. It is therefore different than the conventional understanding of the lenth of a sidereal year, namely, 365.256363 units of 24 hour days.

familiar convention of twenty four hour days. A popular annual fire ritual was continued for three hundred and sixty sidereal days starting with the solstice and its period is called the Savana year. The choice of a 360 day long year is explained ahead in this chapter. Vedic seers expressed their mathematical ingenuity in adjusting the difference in the length between the Savana and the Solar year in their guidelines for the fire rituals extending beyond a year. The two halves of a year, called the *Ayanas*, separated by the solstices carry forward the idea of the bright and the dark *Pakshas* (halves) from the lunar cycles. Twenty four Pakshas in a lunar year correspond to twenty four Ayanas in a *Brihaspati* cycle. Brihaspati or Jupiter, the largest planet in the solar system, takes twelve years to move through the twenty seven Nakshatras. A sixty year cycle combines the movement of Jupiter and Saturn the largest planets of the solar system.

Hints about Precession

Time units higher than sixty years unite to the precession of the earth's poles in an abstract way. An Ayana, one half of a sidereal year, separated by the winter and summer solstices, is called a *Deva* day or night. The daytime of a deva coincides with the night of an *Asura* and vice versa, Three hundred and sixty Deva days constitute a Deva year. The Savana ritual is a miniature imitation of a Deva year. The word *Ayana* in Sanskrit means Precession. The Deva year is a base unit of precession measures. Twelve thousand Deva years create one *Chaturyuga* unit of time. A Chaturyuga consists of 4.32 million sidereal years. A Chaturyuga period is split into four unequal parts. The word *chatur* means four and the word *Yuga* means an era. The following table summarizes the unequal division of a Chaturyuga

Table 2 : The structure of the Chatur-yuga system

	Deva Years	**Sidereal Years**
Krita–Yuga Period		
Dawn	400	144,000
Krita–Yuga	4,000	1,440,000
Twilight	400	144,000

Subtotal	**4,800**	**1,728,000**
Treta–Yuga Period		
Dawn	300	108,000
Treta–Yuga	3,000	1,080,000
Twilight	300	108,000
Subtotal	**3,600**	**1,296,000**
Dvapara–Yuga Period		
Dawn	200	72,000
Dvapara–Yuga	2,000	720,000
Twilight	200	72,000
Subtotal	**2,400**	**864,000**
Kali–Yuga Period		
Dawn	100	36,000
Kali–Yuga	1,000	360,000
Twilight	100	36,000
Subtotal	**1,200**	**432,000**
Total	**12,000**	**4,320,000**

Colonial scholars ridiculed the concept of the *Chaturyuga* system and ascribed it to the obsession of the Indian civilization with mega units of time. The cyclical insight of time of the Vedic seers is a cryptic expression of cosmic scale progressions which modern astronomers are now discovering. Starting with the work Surya Siddhanta, Indian astronomers were fusing parts of the cyclical concept of time into mathematical models. For example, the Surya Siddhanta, divides the Kali Yuga era shown in Table 2 above into four equal parts of 108,000 years and assumes an alignment of planets at the start of each unit of 108,000 years. Aryabhata, a famous astronomer, while presenting concise mathematical formulae for predicting the positions of planets takes Feb 18, 3102 BCE as the point of the last alignment. Feb 18, 3102 is accepted by the Pundits as the start of the current Kali Yuga cycle. Colonial astronomers looked for a perfect alignment of planets on that date and finding none questioned the credibility of the Chaturyuga system. They also assumed that the reference to 4,320,000 units is to tropical years. However, the tropical year is 365.4 days a few hours shorter than a sidereal year of 366.25 days (measured in diurnal rotations of the Earth). The difference between these two numbers alters the accuracy of the Chaturyuga system as a mathematical construct of the precession phenomenon. The following figure, credited to K.D Abhyankar, shows the planetary alignments on two different days around BCE 3102 (Abhyankar K. , 1996). Similar configurations are seen several centuries later but a later time frame does not support references to the equinox and eclipse phenomenon found in the related text. Astronomers can plot the positions of the Sun, Moon, and the planets in the sky, for any day, millenia in the past. In his analysis, K.D Abhyankar discusses the assembly of planets in question in BCE 3104 as well as the notion of the Sun and the Moon being at the Vedic zero point on Feb 18, BCE 3102. One possibility is that astronomers such as Aryabhata[11] were attempting to reestablish the use of astronomical models that may have been discontinued in previous centuries. Modern researchers in India are pursuing an analysis of astronomical conjunctions constrained by the guidelines set forth for Vedic rituals. Their analysis has already resolved various issues raised by the misunderstanding of colonial scholars, freeing intellectual efforts to hone in on deciphering the meaning of the large units of time in the Surya Siddhanta.

Decrypting ideas on Precession

The Chaturyuga period may appear arbitrary in the absence of understanding of the intent behind its design; as with numerous other segments of Vedic knowledge, it conveys knowledge of precession in a cryptic fashion. Neither the number 4,320,000, the duration of a Chaturyuga, nor any of its multipliers make the phenomenon precession explicitly clear. The clue is hidden in the formation of higher units of time, namely, the Manu, the Kalpa, and the Brahma unit of time. A translation (Burgess, 1998) of Surya Siddhanta passage related to units of time proceeds as follows:

> *And sixty nadis make a sidereal day and night. Of thirty of these sidereal days is composed a month; a civil (savana) month consists of as many sunrises;*

---

[11] the earliest known astronomer who attempted to correlate the start date of the Panchang calendar system to an astronomical occurrence in the sky

A lunar month, of as many lunar days (tithi); a solar (saura) month is determined by the entrance of the Sun into a sign of the zodiac; twelve months make a year. This is called a day of the gods.

The day and night of the gods and of the demons are mutually opposed to one another. Six times sixty of them are a year of the gods, and likewise to the demons.

Twelve thousands of these divine years are denominated as a chatur-yuga ; of ten-thousand times four hundred and thirty two solar years is composed the chatur-yuga with its dawn and twilight.

The tenth part of a chatur-yuga, multiplied successively by four, three, two, and one, gives the length of the krita and the other yugas: the sixth part of each belongs to its dawn and twilight.

One and seventy chatur-yugas make a manu; at its end is a twilight which has the number of years of a krita-yuga, and which is a deluge.

In a kalpa are reckoned fourteen manus with their respective twilights; at the commencement of the kalpa is a fifteenth dawn, having the length of a krita-yuga.

The kalpa, thus composed of a thousand chatur-yugas, and which brings about the destruction of all that exists, is a day of Brahma; his night is of the same length.

His extreme age is a hundred, according to this valuation of a day and a night. The half of his life is past; of the remainder, this is the first kalpa.

> *And of this kalpa, six manus are past, with their respective twilights; and of the Manu son of Vivasvat, twenty seven chatur-yugas are past;*

> *Of the present, the twenty eighth chatur-yuga, this krita-yuga is past…*

The Manu unit of time includes many Chaturyuga units while a Kalpa unit of time is derived by multiplying the Chaturyuga period by 1000. Each Kalpa is comprised of 14 Manus, which consist of 71 Chaturyugas. This system yields extraneous years, which, in a cryptic way, are distributed as transition periods. The following table shows the arithmetic of this model:

Table 3 Manu and Kalpa Units of Time

Name	*equivalent to*	*Sidereal Years*
Chaturyuga		4,320,000
Manu	71 Chaturyugas	306,720,000
	+1 Transition	1,728,000
	= Total	308,448,000
Kalpa	14 manus	4,318,272,000
	+1 Transition	1,728,000
	= Total	4,320,000,000

The number of solar/sidereal days (diurnal rotations of the Earth) in a sidereal year is 366.2563795. The fractional solar day required to complete one sidereal year is 0.2563795 days. This fractional contribution turns into an integer, 7, over a period of 10,000 sidereal years. A common Vedic approach for handling arithmetic operations between two decimal numbers is to first convert them into full integers through multiplication with a common number. The Surya Siddhanta text abounds in examples of this approach. The number 10,000, which brings the fractional day count to a whole number, when divided by seven, yields the number, 1428 4/7. This number represents the number of sidereal years required to turn the fractional 0.2563795 day contribution into a full sidereal day. The number 1428 4/7 is the foundation for the higher units of time measures. Manu and Kalpa units can be seen to be the products of the integer **1428** and the associated fraction **4/7** in the following way

> *The number representing a transition period, namely, 1,728,000 is a product of the number of sidereal years in Kali Yuga (432,000), the fraction* ***4/7*** *and the number seven. Similarly the number of years in all the fourteen Manus, namely, 4,318,272,000 is a product of the number of sidereal years in Kali Yuga (432,000) multiplied by the number* ***1428*** *and the number Seven.*

The number seven here is used as a multiplier to convert the fractional number -- the product of the Kaliyuga years 432,000 with 4/7 -- into a whole integer. Incrementing this product by 7 also requires incrementing the product created with the whole number 1428. The number 7 plays an important role in expressing precession in the design of higher order time units. Rishi *Dirghatama* in the Rig Veda, Mandala 1, section 164 talks of the importance of the number 7:

> *Twelve spokes, one wheel, navels three. Who can comprehend this? On it are placed together three hundred and sixty like pegs. They shake not in the least. (Stanza 48)*

> *A seven-named horse does draw this three-navel wheel... Seven steeds draw the seven-wheeled chariot... Wise poets have spun a seven-strand tale around this heavenly calf, the Sun. (Stanza 1-5)*

Glenn R Smith provides the analogy of a clock-like mechanism for the phrase "navel three" found in the above translation (Smith). A sidereal year consisting of 366.2563795 days may be split into three motions, namely, a 360 day motion, a 6 day motion and a 0.2563795 motion. The three navels can be understood to refer to each of these three motions. One can visualize three gears with the ratio of the above numbers, each driving one of the hands of the imaginary clock. A 1:1 ratio gear drives the fast hand of the clock representing the 360 day motion with each of its rotations. A 60:1 ratio gear drives the slow hand of the clock representing the six day motion by a six degree movement with each rotation of the hand of the clock. An even higher ratio gear drives the slowest hand of the clock representing 0.2563795 day motion, moving a corresponding fraction of a degree with each rotation of the fast hand of the clock.

With a one degree movement of the clock's hands representing a sidereal day, the clock, following one rotation of its slow hand, shows the time as one sidereal year. The clock's slow hand completes one circle in 60 sidereal years, and its slowest hand completes one circle in 1404 years. When the slowest hand has completed one full rotation, the total number of rotations of all the hands of clock is 1428. The slowest hand contributes seven of the 10,000 combined rotations of all the hands. The numbers 10,000, 1428 and 7 as described in the earlier paragraph are seen in the motions of this mythical clock.

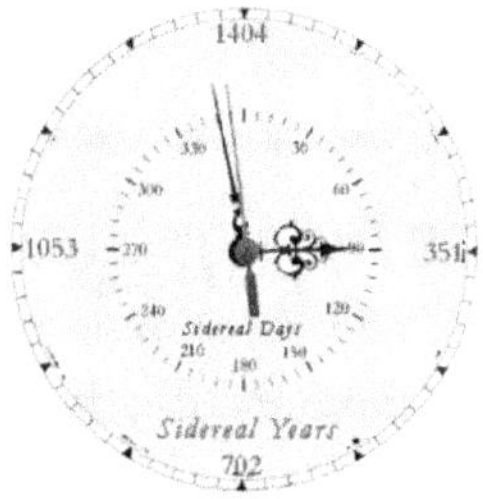

Figure 7 A mystic precession clock: the small hand completes 7 rotations in 10,000 sidereal years

The Chaturyuga with 4,320,000 years is an abstraction illustrating the difference between the lengths of the sidereal year and the solar (equinoctial) year. Kaliyuga is a tenth of a Chaturyuga, while the number 432,000 is the common multiplier for the total number of years in fourteen Manus and in the length of the transition period. The number of precession degrees in a fourteenth of a Kalpa, as per the rate of precession[12] in the Surya Siddhanta, is the number 4,320,000. This represents 12,000 complete precession cycles. The number 12,000 is also the number of *Deva years* in a Chaturyuga unit. The play of the numbers in the Deva, Manu, and Kalpa units of time is the result of a cognition of the precession phenomenon.

The Deva, Manu, and Kalpa units can also be analyzed in in a slightly different way. It takes $71\frac{3}{7} = 71.42857$ sidereal years to observe a 1 degree movement in precession, as per the Surya Siddhanta. The significance of this number is

[12] Surya Siddhanta states the rate of precession per year to be 50".4 = 0°.014 = $\frac{7}{500}$ degrees.

apparent in the analysis of the Manu and Kalpa units of time. 1/14th of a Kalpa is 71.42857 Chaturyugas. A single Manu's contribution to this fourteenth of a Kalpa period of time, including this Manu's transition period, is 71.4 Chaturyugas. The single transition period at the end of a Kalpa is common to all Manus; when distributed equally among the fourteen Manus, it equals 0.02857 Chaturyugas. Thus, the number of precession degrees in the fourteenth of a Kalpa described above is $(71\frac{3}{7}) \times 4{,}320{,}000 \times 0°.014 = 4{,}320{,}000$. This corresponds to 12,000 precession cycles in the sidereal system, and thus connects back to the same figure introduced earlier: 12,000 is number of Deva years in a Chaturyuga and the length of a Chaturyuga in sidereal years.

Precession in stories

Precession occurs around the fixed poles of the sidereal system. The common translation of the Sanskrit word *Dhruva* is "fixed," but there is no bright star in the sky at the position of the fixed sidereal pole. Colloquially however, Dhruva is referred to as the Pole star; the Surya Siddhanta uses the word Dhruva to reference the sidereal pole. This reference may stem from the literature of the Puranas which, in the format of stories and anthropomorphic representations of cosmic principles, conveys complex concepts in the Vedas to the average citizen. A grandmother telling a bedtime story could, for example, point to a fixed star in the sky while relating the story of Dhruva (see below).

The celestial pole is observationally fixed over a human lifetime, shifting a fraction of a degree. That the layman might identify Dhruva as fixed relative to its surrounding, shifting stars was indeed justifiable. The stories of the Puranas hint at concepts more scientifically backed and explained in the other forms of Vedic literature, but intellectual pursuit of these ideas conveyed in stories may have diminished in the centuries following the Vedic period. The following text of the story of Dhruva story shows an example of the similarities between the stories and scientific ideas of Vedic astronomy:

> *Vishnu grants Dhruva a permanent deathless place far above the Saptarishis. This place is the Dhruvamandala. Vishnu describes it as the one around which the*

> Saptarishis *circumambulate in a period of 2600 years while the stars or the Ecliptic take 26000 years to rotate once.*

> *Dhruva married Bhrami, the daughter of Prajapati Simsumara. She had two sons named Kalpa and Vatsara. Vatsara's son was Pushparna. Pushparna was married to Prabha and Dosha. Prabha gave birth to Pratah, Madhyandina and sayamkala. Dosha gave birth to pradosha, nishidhi and vyushti. Vyushti's grandson was Chakshusa.*

This story loads a wealth of information into the names of its characters. An analysis of the *Dhruva* story points to the principle behind precession without invoking mathematics:

> The word *Mandala* refers to a region. *Dhruvamandala* is a region of the sky around which the *Saptarishis* and other stars rotate. *Dhruva* rules over this region.
>
> > The celestial North Pole shifts gradually along an imaginary circle, creating a region in the sky around the sidereal pole around which stars like *Saptarishis* move
>
> The word *Bhrami* means one who keeps moving.
>
> > *Dhruva*, the sidereal pole and *Bhrami*, the celestial poles are inseparable, like a married couple.
>
> *Simsumara* is the name of *Bhrami's* father.
>
> > Astronomical works refer to the ecliptic by the word *Simsumara*
>
> The son of *Dhruva* and *Bhrami* is Kalpa.
>
> > Kalpa refers to the time unit arising from the phenomenon of precession.
>
> The word *Vatsara* means the year.
>
> > *Vatsara* is a part of the Vedic calendar system as described in Chapter 3.
>
> *Vatsara's* grandchildren can be identified, by their names, as the three parts of the day and three parts of the night.
>
> The name of *Vyushti's* great grandson, *Chakshusa* is the name of one of the fourteen Manus.

The three categories of Vedic literature, the *Smriti, Shruti* and *Puranas,* introduce scientific ideas through disparate methodologies. The Puranas refer to the movement of a select star group called the Saptarishis through the twenty seven Nakshatras as a means of indicating the number of centuries that have elapsed. The rotational cycle of the Saptarishis[13] in the Puranas is perplexing. Researchers have interpreted the related text in different ways, but reasonable explanation has largely eluded most efforts. The Saptarishis were in the *Magha* Nakshatra, according to the *Mahabharata* text, at the coronation of the prince *Yudhishtira.* The star Alpha-draconis or Thuban was the closest star to the pole star, centuries prior, and a different set of stars in its vicinity may have served as navigational guides.

Decoding Names of Manus

The Kalpa and Manu names also convey information about the evolution of the cosmos, the solar system, and life on planet Earth. Figure 8 and Figure 9 connect the chronology of the cosmos to the Vedic understanding derived from the names of the Manus and their ages.

We are 1.9 billion years[14] into the current Kalpa, with Puranic literature indicating a Kalpa to be the daytime of Brahma, the cosmic creator. In his waking state, Brahma consciously creates with lucidity, filling the cosmos with the energy of expansion. During the Kalpa night, the Vedic creator slips into deep sleep and dream states, which diminishes coherency in the cosmos. According to modern paleontologists, eukaryotic life forms began to appear on Earth approximately 1.9 billion years ago, distinct from earlier organisms because their cells contained complex structures enclosed within a membrane.

The current Kalpa period will continue for another 2.4 billion years, the Sun likely to be well into a gradual warming phase. Scientists predict that all water from Earth will have evaporated in 2 billion years, and thus, the current Kalpa will end with the Earth devoid of life forms. Evolution and the thriving phase

---

[13] Commonly identified as the seven stars of the Ursa Major Constellation which were used as the navigational stars. These stars do not fit the phenomenon of 2700 year rotations around the pole star

[14] As per the Kalpa, Manu, Chaturyuga definitions, and the customary narration of the position of present day. This narration is a compulsory attribute of Sankalpa done before the start of any Vedic ritual, followed to this day.

of eukaryotic consciousness on Earth map directly to the current Kalpa daytime of Brahma.

Genetic material could be carried across generations for the first time with the arrival of the eukaryotic organism with a nucleus, or membrane bound structure. Cell divisions of these organisms involves the separation of duplicated chromosomes and supports the growth structures needed for sexual reproduction. The cell division in earlier organisms, prokaryotes, occurs without sexual reproduction. The name of the first Manu of this Kalpa is *Swayambhu*, meaning "one created without parents," and is an apt reference to the arrival of eukaryotic life forms on planet Earth

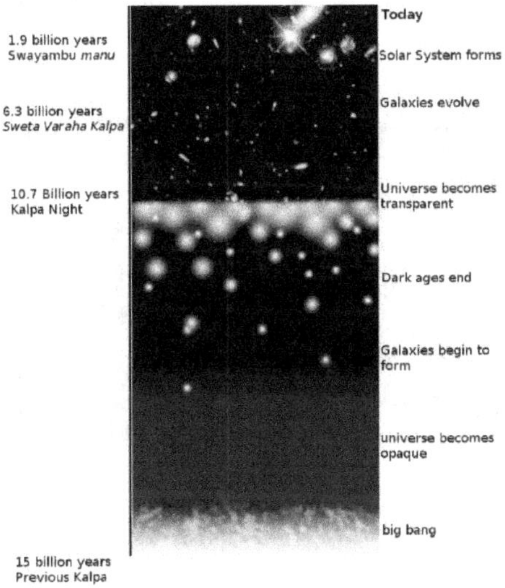

Figure 8 : Kalpas and Modern Cosmology

The names of the subsequent Manus *Swárochisha* ("effulgent/aware"), *Auttami* ("advancement"), *Támasa* ("inert"), *Raivata* ("abundant"), and *Chákshusha* ("sight") coincide with geological and paleontological milestones. Earth's great glaciations occurred within the period of Tamasa Manu, with animals and vertebrates evolving some 500 to 590 million years ago. The post-Cambrian period saw an explosion of species; this period coincides with the Manu named *Raivata* -- abundance. The age of the *Chakshusa* Manu began 430 to 440 million years ago. One theory explaining the rapid increase in diversity of species in the post Cambrian world is the maturation of the organs of sight in life forms. *Chakshusa*, defined over this same time period in the Purana text, is literally defined as "sight." The current Vedic Manu is *Vaivasvata*, whose age began approximately 120 million years ago, corresponding to the Cretaceous period. Twenty seven of the seventy one Chaturyuga cycles have elapsed since the start of the current Manu period. The massive upheavals of Earth, which saw the extinction of dinosaurs, map to the 12th and the 13th Chaturyuga cycles within the current Manu period. The current Manu's age will span another 190 million years.

Additional research on the Manu period indicates that it may be comparable to the rotation of the solar system around the galactic center of the Milky Way. A single revolution of the Sun around this center spans approximately 250 million years. The Sun has then completed close to seven full revolutions since the start of the current Kalpa. Aptly, we are currently in the seventh Manu period within the current Kalpa.

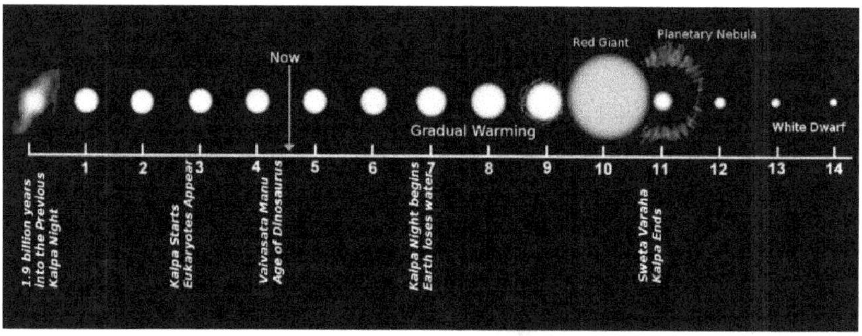

Figure 9 : Life span of the Sun and Manu periods

Vedic civilization gave new meaning to making use of cosmic movement to define measures of time. Human respiration is the shortest of these time measures. The time elapsed over six respirations is defined as one *Vinadi*, with sixty comprising a *Nadi*. Sixty Nadis form one day and a night. Three hundred and sixty Savana days create a Savana year. Vedic mantras link time cycles with anthropomorphic associations, relating the shortest time cycle to man and the longest time cycle to the cosmic being named Brahma. Some elements of modern science postulate that time may just be a production of the human brain, the vast majority of life forms subsisting in an ever-present, moment to moment state of existence. The Surya Siddhanta introduces a similar alternate conception of time (Ravishankar, 2002); indeed, the very concept of time itself[15], loses its charm in a cosmos filled with matter but devoid of intelligence.

[15] There are five different classifications of time in the Vedic tradition

5. Locating the Boundaries

Modern astronomy uses an equatorial coordinate system to determine spatial location. As the name suggests, the system is defined by the celestial equator, its line of zero latitude. The celestial poles function as its 90 degree latitude points.

Vedic astronomers make use of the ecliptic, the path along which the sun -- and, in close proximity, the planets and Earth's moon -- appear to travel in the sky as its zero latitude line and the sidereal poles as its 90 degree latitude points. Understanding the differences between celestial coordinate systems enables an understanding of the historical development of the analysis of planetary motion.

This chapter identifies the Nakshatra boundaries in the modern equatorial coordinate system by mapping them from the Vedic system. It also identifies the coordinates of astronomical objects discussed in this book in the Vedic system, translating the respective longitude/latitude values from the modern coordinate systems. Readers interested in researching correlations between Vedic and modern astronomy will appreciate the examples mapping between the Vedic and modern system. An understanding of the Vedic coordinate system also enables the reader to understand the observational instruments used by Vedic astronomers discussed in Chapter 3.

Astronomical Coordinate Systems:
Astronomers measure the positions of objects in the sky on an imaginary celestial sphere. The celestial poles and the equator are the projections of the Earth's poles and the equator onto this imaginary sphere. A meridian is a circle running from one pole to the other through a point directly overhead to an observer.

Different coordinate systems are efficient in tracking objects in the sky from different vantage points. The Equatorial coordinate system, for instance, is efficient in aligning Earth-based telescopes to the planet's polar axis and equator. Modern stellar catalogues make extensive use of this coordinate system. Telescopes mounted on satellites scan the sky and produce maps of the sky in the galactic coordinate systems. The coordinates of an object in any system contain two components: the object's right ascension (RA) value and its declination (DEC) value.

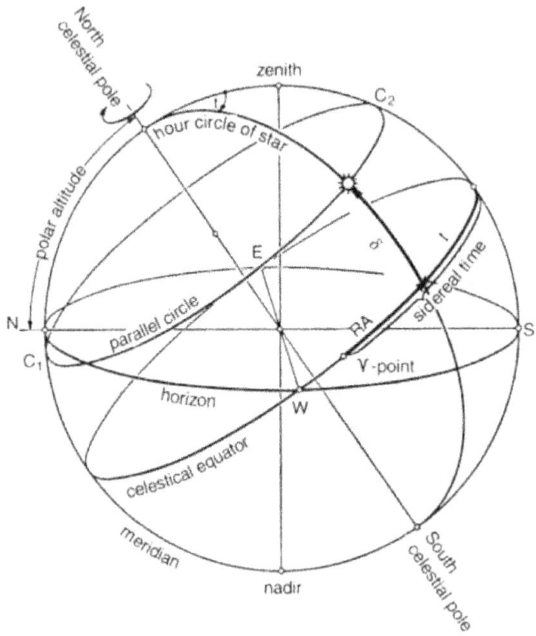

Figure 10 **Spherical coordinate system**

Right Ascension is generally measured in units of time: hours, minutes, and seconds. Historically located at the Vernal Equinox on the celestial equator at sunrise on the first day of spring, the zero-point for RA came to be called the first point of Aries after the discovery that the Vernal Equinox location shifts marginally each year. The total range of right ascension is 24 hours or 360 degrees.

As the RA of the horizon point shifts by 15 degrees in an hour, converting from units of time to degrees is done simply, by a factor of 15 degrees/hour. RA value is analogous to the longitude measurement of the position of a point on the surface of the earth. Similarly, declination value is analogous to the latitude measure of a point on the surface of the earth. It is measured north or south of the celestial equator.

Declination is usually expressed in degrees, minutes of arc, and seconds of arc. The zero points for RA and DEC described can be referenced in methods beyond simply the direction of the Sun at the equinox point. For example, two other widely used systems utilize the direction of the galactic center for their zero points. The three primary coordinate systems in use today are the (a)

Equatorial coordinate system (b) Galactic coordinate system, and (c) Inter-galactic coordinate system. The coordinates of any object in the sky can be easily converted between these systems using spherical trigonometry.

Map of the Precession Circle

The tilt of the Earth's axis with respect to the Earth's orbital plane is responsible for the planet's seasons. Because the equatorial plane intersects Earth's orbit at an angle of 23.5°, the sun, the moon, and the planets do not appear to move along the equatorial circle but rather on the ecliptic. The Earth's North Pole tilts towards the Sun during the summer and away from the Sun in the winter; as a result, seasons in the Earth's southern hemisphere are reversed.

The Earth wobbles slowly like a spinning top with a defined periodicity, a process known as precession. Seasons appear to start on fixed calendar days but in reality drift backwards about a day per century. The two points in the sky that the Earth's North and South poles are aligned with shift in similar fashion. For the past several centuries, the North Pole has pointed towards the region of Polaris, the pole star.

The Earth takes approximately 26,000 years to complete one precession cycle, but this wobble does not affect the angle between the ecliptic and the equatorial planes. Instead, it only affects the point where the ecliptic intersects the equator. This shifting point of intersection makes the tropical year shorter than the sidereal year. So, even as the Earth may return to the same point on its orbital path around the Sun, the position of the Sun as viewed from the Earth will have moved slightly forward in one tropical year with respect to the backdrop of stars. On the other hand, the Sun as seen from Earth aligns against the same backdrop of stars after one sidereal year. Vedic astronomy reflects knowledge of this phenomenon. While the Gregorian calendar is based on tropical years with a length of 365.2422 days, the sidereal year of the Vedic calendar has 365.2564 days.

Figure 11 shows the imaginary circle projected in the sky by the slow shifting motion of Earth's North Pole. The circle shows the location of the celestial North Pole for different eras. The celestial Pole of today is shown close to the highest point of the circle while the celestial Pole at the start of the Kali Era was halfway between the "-2000" and the "-4000" markers on the circle. The center of this circle is the North Pole of the Vedic system. For ages, it has remained fixed against the backdrop of stars in the night sky. The Vedic or the ecliptic North Pole lies at the latitude of N 66° 33" on the J2000[16] coordinate system. No bright stars are found at the ecliptic North Pole. Similarly, the Celestial poles do not match up against any bright stars for several centuries as noted.

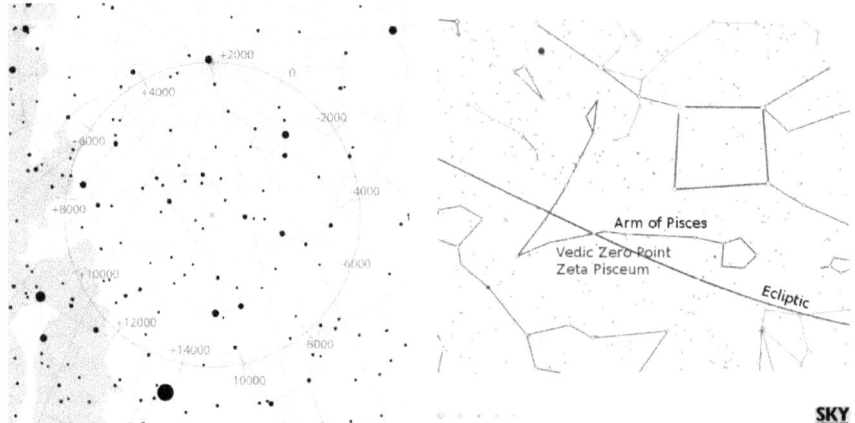

Figure 11 Precession Circle around Sidereal Poles

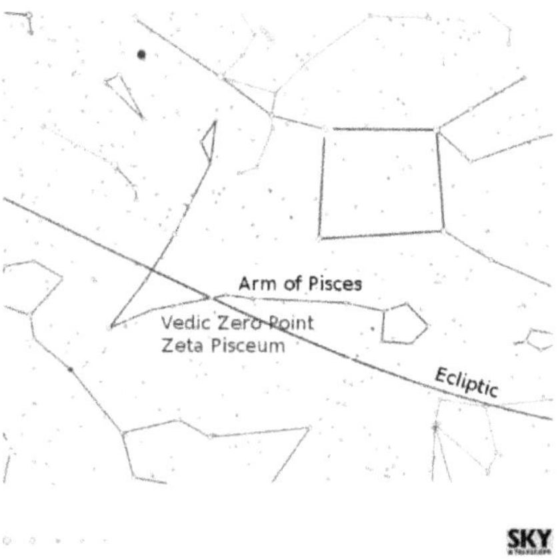

Figure 12 Vedic Zero point at Zeta Piscium

The zero point of the Vedic system

The right ascension of the Vedic system is measured along the ecliptic. Unlike the conventional coordinate system, its zero point (Figure 12) is independent of the changing equinox point.

Ironically, colonial scholars characterized this choice of zero point as arbitrary. Uncomfortable with the idea of unconventional, disparate astronomical discoveries made by the civilizations they sought to oppress, colonial literature frequently quoted an incorrect translation of a Jyotish text referencing an astronomical conjunction associated with the start of an annual Vedic ritual

[16] Modern astronomers adopted the J2000 standards delinking the zero RA point from the position of the equinox. The zero point simply coincided with the position of the equinox in 2000 CE.

and one entirely unrelated with the Vedic definition of right ascension. Sadly, it was a recurring pattern for European translations of Vedic astronomical texts to conform blindly to the flawed timeline colonial assigned to the very existence of the Indian civilization.

Tabulating Nakshatra boundaries

In the equatorial system, Earth's poles project onto the celestial sphere, defining maximum declination values at 90 degrees in the north and south directions. Celestial poles lie perpendicular to Earth's equator.

Just as the North Pole points towards the star Polaris today but will slowly move away from it due to precession, the equatorial projection on the celestial sphere will also change its direction of tilt accordingly. The poles of the Vedic sidereal system are invariant, relative to the precession of the Earth; the Sanskrit term for invariant poles is *Dhruva.* The declination value of a star in the Vedic system is measured along an arc connecting to its invariant poles perpendicular to the ecliptic. Because of its use of the *Dhruva* or fixed poles, the Vedic system is fundamentally a polar coordinate system. Declination in the Vedic system is termed *Viksepa* in Sanskrit while right ascension is called *Apakrama.*

Twenty eight specific stars, one in each Nakshatra, were used as observational anchor points. These stars were termed *Yoga Tara* and helped Vedic calendar makers make periodic rectifications over the centuries to account for the shifting equinox point to the positions of the Sun and planets as predicted by the formulas presented in astronomic treatises.

This chapter tabulates the RA coordinates for the *Nakshatra* boundaries, mapping them from Vedic coordinates to the conventional equatorial coordinate system. Table 4 includes the RA of the representative astronomical objects discussed further in Chapter 8. Each *Nakshatra* is represented by an arc measuring 13 degrees 20 min along the ecliptic. A longitudinal line from the *Yoga Tara* of a *Nakshatra* projects inside its own arc along the ecliptic. The longitudinal value of a *Yoga Tara* is not measured from the RA zero point but rather is measured relative to the *Nakshatra* arc. The declination of a Yoga Tara star, however, is measured from the ecliptic towards the sidereal poles, North or South.

The following table shows the starting RA for each Nakshatra using the J2000 ecliptic in the second column. The fourth column of the table shows RA coordinates of representative astronomical objects for each Nakshatra. The third column lists the name of the representative object. The text Surya Siddhanta mentions a Nakshatra system consisting of both half-sized and one-and-one-half sized Nakshatra arcs. The fifth column shows the starting RA coordinates Nakshatras in this alternate arc size system.

Table 4 : Coordinates of Nakshatra arcs and representative objects

Nakshatra	Arc start °	Object	Obj RA	Alt Arc start
Krittika	3.3	Pleiades	4	2.8
Rohini	4.2	Aldebaran	4.6	3.7
Mrigashira	5.1	Orion Molecular Clouds	5.5	5.1
Ardra	5.9	Betelgeuse	5.9	5.9
Punarvasu	6.8	Abel 21	7.5	6.4
Pushya	7.7	Cancri 55	8.5	7.7
Ashlesha	8.6	Epsilon Hydare	8.8	8.6
Magha	9.5	Leo I Galaxy	10	9
Falguni – Purva	10.4	Leo Cluster	11.2	9.9
Falguni – Uttara	11.3	Whirlpool Galaxy	11.7	10.8
Hasta	12.2	NGC 4552	12.2	12.2
Chitra	13.1	MTN 69	13.2	13.1
Nishti/Swathi	13.9	Galactic Wind	14-15	13.9
Vishaka	14.8	galactic central bar end	15	14.4
Anuradha	15.7	Great attractor	16.9*	15.7
Jyeshta	16.6	Galactic Central region	17	16.6
Mula	17.5	Galactic Black Hole	17.7	17
Ashada – Purva	18.4	RGSC 1,2,3	18-19	17.9
Ashada – Uttara	19.3	galactic central bar end	19-20	18.8
Shravana	20.2	Cygnus X1	20.9	20.2
Shravishta	21.1	Messier 15	21.9	21.1
Shatabhishaj	21.9	NGC 7492	22.8	21.9
Proshtapada (P)	22.8	NGC 7331	22.6	22.4
Proshatapada (U)	23.7	Andromeda α	0.01	23.3
Revati	0.6	Alpha Cephei	0.9	0.6
Ashwini	1.5	Beta Arietis	2.3	1.5

6. *Refining the View*

Galileo's telescope put an end to the debate about the nature of the hazy band of light in the night sky known as the Milky Way. The grand scale of vision granted by the telescope cleared the ground for mapping the structure of our galaxy.

Until the dawn of the twentieth century, modern astronomy largely placed the Sun at the center of the Milky Way. But the Vedic Rishis had cognized the center in the direction of the *Mula* Nakshatra. In 1920, when astronomer Harlow Shapely postulated that the Sun must be far removed from the center of the galaxy, the idea of a distant center started taking concrete shape. The bright spiral disc shape we now recognize as the shape of the Milky Way was gradually observed and pieced together by scientists over the course of decades. By scanning the sky across various wavelengths of light, vision beyond objects obscuring visible light became possible. However, the visualization of the final barred spiral structure of the Milky Way still relies on inferences from the structure of neighboring galaxies.

This chapter maps the boundaries of Nakshatras onto the all sky map format, commonly seen in data provided by modern sky scanning satellites. Nakshatra boundaries are not obvious without the use of maps. Moreover, this chapter explains how Nakshatra names reflect aspects of the structure of the Milky Way.

The Sun and the Earth sit in a habitable zone of the Milky Way galaxy, about 16-98 light years above the central disc. The Galactic disc, dark to our telescopes due to interstellar dust, is characterized by a high density of stars and a high rate of star formation. The Sun sits in a minor arm of the Milky Way, slightly above the chaotic plane of this disc, the solar system at a distance of more than 27,000 light years from the galactic core. The Orion arm, as it is termed, contains relatively lesser star formation activity. Within this arm, the Sun exists in a quiet region defined as a local bubble[17].

Astronomers have mapped the majority of stars in close proximity to the Sun; many of these lie scattered in the shape of a small belt within the Orion arm. This belt region can be viewed as a linear arrangement of stars in the night sky.

---

[17] The inside of the bubble consisting of higher density galactic gas compared to outside gas gives it the name

Away from the Sun, in the direction opposite the center of the Galactic core, lies a major arm of the Milky Way: the Perseus arm. 6500 light years away from the Sun, the Perseus arm has more active star in multiple regions. Astronomers use the galactic coordinate system to visualize the positions of the stars and the galactic arms with respect to the Sun. In this particular coordinate system, the direction towards the galactic center from the Sun is defined as zero degrees.

The Sun lies at the origin of the galactic coordinate system (Figure 13). Galactic longitude measures the angle between the direction from the Sun to a star and the direction from the sun to the center of the galaxy while galactic latitude measures the angle of the object above the galactic plane. An all sky map uses these galactic coordinates to portray a celestial object position. While terrestrial telescopes can be easily oriented using equatorial coordinates, galactic coordinates provide the most convenient method to orient telescopes mounted on satellites. Conversion tools make translating between the two systems trivial.

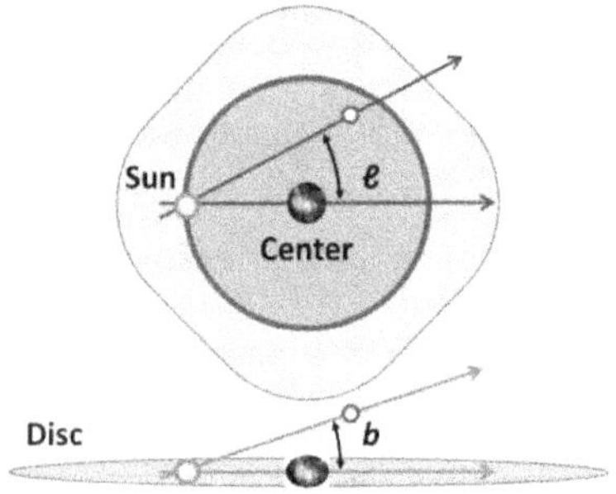

Figure 13 Top and side views of the galaxy illustrating galactic longitude and latitude of a star

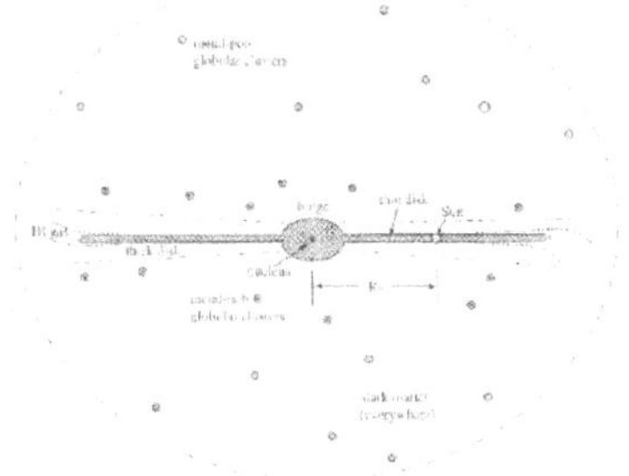

Figure 14 A side view of the Galactic disc, showing a bulge at the center of the galaxy, the galactic disc and the galactic halo

Figure 15 Center of the Milky Way rising in Black Rock Desert in Nevada. The thick band of stars and dusty area is the galactic disc

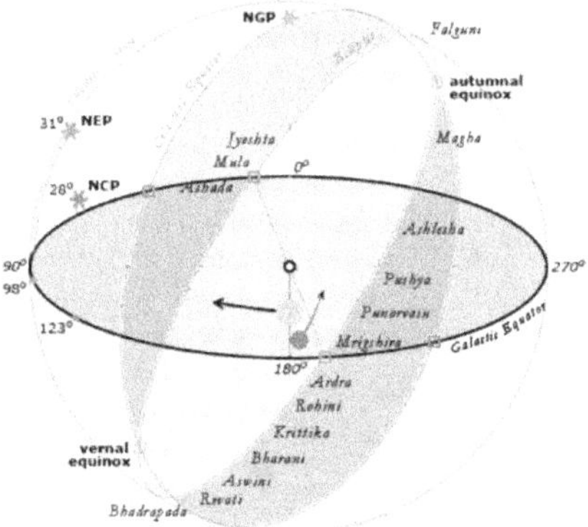

Figure 16 Three spherical coordinates illustrating the tilt among the galactic plane, the ecliptic plane and the equatorial plane

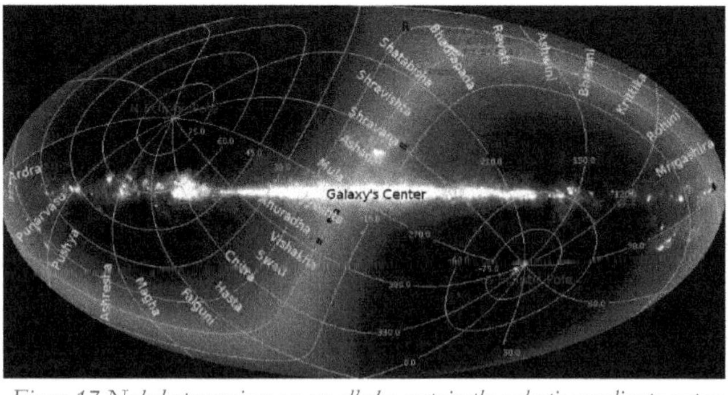

Figure 17 Nakshatra regions on an all sky map in the galactic coordinate system

The vast majority of stars in the sky are dimmer than an absolute magnitude[18] of 8.5. Scientists draw this conclusion from the fact that there are approximately 208 stars brighter than absolute magnitude 8.5 within a sphere of a 49 light year radius from our Sun, but this count drops to 64 within a 16 light year radius, even if stars of less than magnitude of 8.5 are included. The Sun, as do all other resolvable stars in the sky, revolves around the center of the Milky Way galaxy.

The current movement of the Sun is in the general direction of the star Vega in the constellation Hercules. This solar motion occurs at an angle of 60 sky degrees from the direction of the galactic center. The Sun also oscillates up and down relative to the Galactic plane, approximately 2.7 times per orbit. These oscillations have been postulated to coincide with mass extinction periods (James, 1984) on Earth. The density of stars within the Milky Way increases as one moves closer to the galactic disc, as shown in Figure 14. Astronomers have mapped the location of a multitude of stars within a distance of 1000 light years from the Sun.

Seen from the Earth, a part of the flattened disk of our Galaxy appears as a band of light across the heavens, an image readily evoked upon hearing the words "Milky Way". But this band of light is but a small portion of the Galaxy's disk. In the opposite direction from the band, another small view of our galaxy's disk can be found. Because of the tilt of the solar system relative to the galaxy, the disc is not distinctly visible in regions other than these two directions. During the winter months one can readily observe the path of the planets in the sky tilted 60 degrees away from the band of the Milky Way. The orbital plane of the Earth around the Sun is inclined at an angle of about 60 degrees to the orbital plane of the Sun around the core of the Galaxy. Figure 15, shows a night sky view of the Milky Way towards the Constellation Sagittarius, seen from the Black Rock Desert in Nevada.

The center of the Galaxy is obscured by thick interstellar clouds of gas and dust between the Earth and the center of the galaxy. The galactic bulge is observable as an ellipsoid of stars above and below the Galactic plane, a darker band of dust and clouds. Star population close to the center of the galaxy is

[18] Star magnitudes count backward. The brightest ones are of the first magnitude and those not so bright are called the second magnitude and so on.

very dense -- hundreds of thousands of stars per cubic parsec[19]. 100 parsecs away from the galactic center, the star density in the galactic core drops to approximately one star per 100 parsecs. The star density is one per cubic parsec in the neighborhood[20] of the Sun.

Exacerbated star density near the galactic center is a function of gravity. Despite this elevated density, most stars remain invisible due to interstellar dust. The majority of energy from the galactic core received on Earth is in the infrared spectrum; as a result, infrared telescopes are commonly used to study the galactic core. Analysis of emitted radiation suggests that an accretion disk ranging on the order of 10 parsecs exists around the center of the galaxy, (further detailed in Chapter 8).

Without the 60° tilt of our solar system, ostensibly every view of the celestial equator in the night sky would show a portion of the galactic disc. But due to the obliquity of our local environment, the galactic disc is only distinctly identifiable near the constellations Sagittarius and Gemini where it crosses the celestial equator. By the same token, distant, extra-galactic objects are obscured by the galactic center. A telescope pointing toward the disc in the directions of Sagittarius cannot capture objects lying beyond the Milky Way, but a telescope pointing through the disc or away from the disk in, say the directions of Leo or Pisces, has far clearer views of distant objects in the universe.

The sun, the moon, and the planets move in the sky along the ecliptic, close to the celestial equator. Sometimes, they lie in a Nakshatra arc containing the galactic disc and at other times in a Nakshatra arc positioned against a background of the most distant visible objects in the universe. Since the Sun is positioned multiple light years above the galactic plane, a viewing angle through the galactic disc in the Proshtapada Nakshatra arc and away from the galactic disc in the Phalguni Nakshatra arc is created. Translations of phrases in the mantras and the names of the two Nakshatras, may hint at this feature of the solar system.

Figure 16 presents the relationship of the galactic coordinate system to the ecliptic and equatorial coordinate systems. The Galactic equator, the Celestial equator and the ecliptic are represented by circles in the figure. The intersection points between the celestial and the ecliptic circles are the points of equinoxes.

[19] A parsec is 3.26 light years
[20] The Sun is 8-9 kpc distance from the galactic center

Nakshatras from *Falguni* to *Proshtapada*, are noted along the front face of the ecliptic and three Nakshatras are shown in the vicinity of the galactic center on the opposite side. The ball representing the Earth is moving around the ball representing the Sun. The curved arrow around Earth points to the orbital direction of the Earth around the Sun. The Nakshatra in which the Sun is seen in the sky is on the ecliptic circle, on the far side from the Earth. When the Earth moves around the Sun in the stellar plane, the Sun appears to move into different Nakshatras. The line connecting the Sun to the galactic center is the zero degree longitude of the galactic coordinate system. The 180 degree longitude of the system is the extension of the same line in the opposite direction. The Sun itself moves along the galactic plane in the direction indicated by the arc arrow attached to the Sun in the figure.

Telescopes such as the Hubble, strategically positioned around Earth, scan the sky in various wavelengths. Astronomers generate maps from this data, using the galactic coordinate system in the all sky map format. Figure 17 shows the position of the ecliptic in an all sky map presented in the galactic coordinate system. The galactic center lies in the middle of this figure, in the region of the *Mula* Nakshatra. The highest and lowest declinations, in galactic coordinates, of the ecliptic fall within the *Proshtapada* and *Falguni* Nakshatra regions respectively.

The horizontal region of brightness passing through the galactic center represents the galactic disc. The galactic anti center is represented at the two ends of this horizontal line. The galactic anti center lies at the boundary between *Mrigashira* and *Ardra* Nakshatras. The names of the Nakshatras are printed next to the ecliptic and oriented along the lines connecting the two ecliptic poles and cross the ecliptic at a 90 degree angle. Representative lines connect the two ecliptic poles shown in the figure as curving lines. The boundary between two neighboring Nakshatras is one such curving line. The galactic coordinates of the ecliptic north and south poles are (70, 27) and (250, 27). The Galactic 90 degree longitude is seen in *Proshtapada* (Purva) and the Galactic 270 degree longitude in *Phalguni* (Uttara). The arcs of the ecliptic lie furthest from the galactic disc in these two Nakshatras. In the night sky, in the sidereal coordinate system, the highest and lowest declinations of the obliquely oriented galactic disc are reversed and lie in the Proshtapada and Phalguni Nakshatras respectively.

Table 5 Boundaries of Nakshatras on the galactic disc

	Galactic Longitude	Sidereal Latitude
Ashwini	109	58
Bharani	117	55
Krittika	127	49
Rohini	142	38
Mrigashira	162	22
Ardra	186	0.2
Punarvasu	213	-23
Tishya	233	-39
Ashresha	246	-48
Magha	257	-55
Falguni – Purva	264	-58
Falguni – Uttara	274	-60
Hasta	278	-60
Chitra	285	-59
Nishtya	292	-57
Vishakha	302	-52
Anuradha	314	-44
Jyeshta	330	-31
Mula	354	-12
Ashada - Purva	20	12
Ashada - Uttara	44	32
Shravana	60	45
Shravishta	72	52
Shathabhishaj	81	57
Proshtapada - Purva	88	59
Proshtapada - Uttara	95	61
Revati	101	59

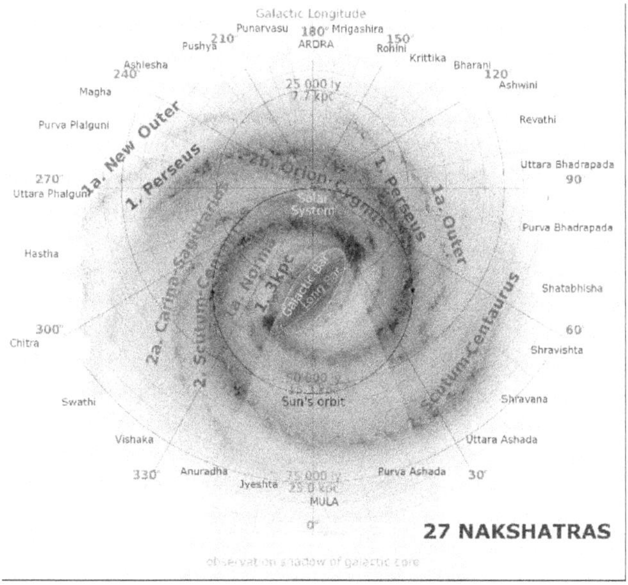

Figure 18 Nakshtra positions marked on the map of Milky way Galaxy

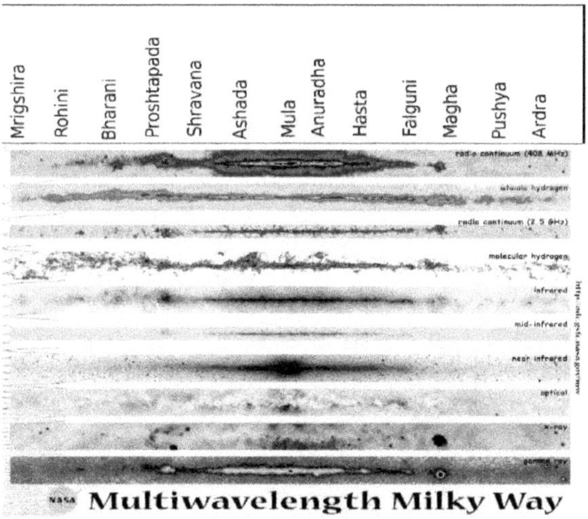

Figure 19 Spectrum mapping along galactic disc in different frequencies (top to bottom) - Radio (0.4 GHz), Atomic Hydrogen, Radio (2.7GHz), Molecular Hydrogen, Infrared, mid infrared, near infrared, optical, x-ray, gamma ray.

Nakshatra arc boundaries on the galactic disc are determined through calculating the longitude of the boundary points in the galactic coordinate system by keeping the galactic latitude zero. The sidereal latitudes for the boundary points on the galactic disc are then determined using the coordinate transformation formula and the galactic extinction calculator. Table 5 shows these coordinate points. Figure 18 Nakshtra positions marked on the map of Milky way Galaxy using the values in this table. Figure 19 shows the locations of Nakshatras on maps of the galactic disc. In the next chapter we look for specific astronomical phenomenon, structures, or objects that match the cryptic messages behind the ideas of Nakshatras and their Devatas.

7. Decoding Devata Names

Chapter 5 introduced 27 structures in the cosmos, some within the Milky Way galaxy and others beyond. Each of these unique structures draws attention to the nature of the Nakashtra region in the sky in which it appears. This chapter provides a general understanding of Devatas and delves into the meanings behind ancient Nakshatra mantras. The next chapter highlights the salient features of astronomical objects, each of which parallel the Vedic understanding of the Nakshatra and its Devatas.

Surya Siddhanta, the ancient treatise of Vedic astronomy, contains a list of Nakshatras that partition the ecliptic in an easily observable manner, one that facilitates study of the movements of planets. These partitions have been used to measure many units of time or epochs -- day, month, year, a cycle of 12 years, a cycle of 60 years, and a cycle of 12,000 years among others. Surya Siddhanta uses three distinct terms: "Graha"-- a luminary that moves along the ecliptic, "Tara" -- a stationary luminary, and "Nakshatra" -- a division of the sky. However, colloquially, "Tara" and Nakshatra are used interchangeably even though a Nakshatra more closely resembles a traditional constellation.

Star gazers, in medieval times imagined a form or a figure connecting a set of prominent stars in a region of the sky and named the constellation after that figure. For example the shape of a scorpion can be visualized in the imaginary lines connecting prominent stars in the constellation Scorpio. However, a Nakshatra's name and the characteristics of its Devata connect it to an astronomical phenomenon that dominates the region via proximity and uniqueness. Advances in modern astronomy aid comprehension of the logic behind Nakshatra names and Devata assignments. Contemporary images of Devatas which evolved through historic times convey only a few of the myriad ideas from the Vedic period. We therefore look for additional ideas in Vedic texts such as the Nakshatra *Sukta* mantras and the *Nirukta* of *Yaskacharya.*

Rishis, Vedic seers, conceptualized the unseen energies that drive the cosmos as Devatas. The oldest attempt at analyzing this is seen in the text "Aitareya Brahmana" (Keith, 1920) which explains the logic behind the selection of mantras relevant to popular Vedic rituals. During historical times, Shankaracharya provided a commentary on the text "*Brihadaranyaka Upanishad*"

(Madhavananda, 1950) in addition to the logic behind the classification of Devatas into 33 categories. He expounded the ancient knowledge to a post-Buddhist society. During the last century, Sri Aurobindo (Sri Aurobindo, 1971) and Swami Dayanand Saraswati (Saraswati, 2011) wrote commentaries on Vedic texts, explaining Devata forces in the context of modern science.

The use of the *Nirukta* approach has been considered essential in validating new conclusions on Vedic texts pertaining to the science of formation of names from word roots. It, like the five other treatises, is a limb of the Vedas, designed to facilitate an interpretation of the Vedic texts. Nirukta, in its first chapter, emphasizes the importance of the scientific spirit of inquiry. It introduces a critic named Kautsa, who not only questions the authority of the Vedas, but actually maintains that Vedic stanzas are meaningless, adducing several arguments in support of his assertion. The author of the Nirukta then establishes his own belief that the Vedic Mantras have been cognized and require care in interpretation. Lakshman Swaroop, who provided a popular English translation of the Nirukta text, writes

"*The epoch of Yaska was an age of remarkable literary activity. There seems to be a general striving towards truth in all the departments of human knowledge.*"

He proceeds to write about the critic Kausta mentioned in Nirukta:

"*The reproduction of the Kautsa controversy indicates on the one hand, that not only Yaska was endowed with a rationalistic spirit, and was free from bigoted fanaticism, but also that it was possible to carry on such discussions with tolerance in that period of remote antiquity*"

The Nakshatra mantras used in the interpretations in this chapter belong to the Taitriya Shaka of the Yajur Veda which are still being used by Pundits. This recension has survived intact through the Gurukul system of education discussed earlier. It is likely that other mantras related to Nakshatras were present in the lost recensions, potentially very useful in astronomical interpretations. Each Nakshatra's Mantra refers to a Devata. Yaskacharya, the author of the Nirukta explains different ways to analyze Devata names by quoting examples from the Vedas.

60 years ago, Paul-Emile Dumont published a word for word translation of these Nakshatra mantras, but unfortunately, such translations of Vedic texts have significant limitations. The first limitation arises from the use of

intonation, an integral part of Vedic mantras. Intonations have generally been preserved through the oral tradition and can alter the meaning of a sentence just as punctuation marks do in the English language. Information carried through intonation is the first to be lost in any word to word translation.

The second limitation relates to the choice of a particular word definition, amongst alternatives, based on textual context. According to the Vedic understanding of the cosmos, Devatas are fundamental impulses whose influence is seen on many fronts, from the micro to the macro level. The context of a mantra must be understood before choosing from among potential meaning choices. English translations during the colonial period preferred a meteorological context for Devatas: for instance, Indra, Rudra, Soma, and Vayu. But at a micro level, the text related to these Devatas can be understood from a physiological and neurological context as well. The possibility of an astronomical context is also feasible based on the latest theories on the evolution of the cosmos. This chapter quotes Paul-Emile Dumont's translations and also gives alternate interpretations that fit better in an astronomical sense. Alternate word meanings can be found in the Monier Williams dictionary.

Understanding the relative position of the Taitriya Shaka and Nirukta in the vast body of Vedic knowledge is important. The main body of Vedic literature is split, based on the types of Mantras they hold (a) Riks (b) Yajus (c) Samans and (d) Atharva. Some use the example of the field of music to explain this fourfold division. Stringed instruments produce music. If the science of vibrating strings and the different frequencies thus produced compares to the Riks, the engineering of musical instruments are the Yajus, the art of using a stringed instrument are the Samans, and the study of the effect of music on human moods are the Atharva mantras. The Riks, or the Rig Veda, is the logical place to look for the science of Devatas. The Nirukta quotes mostly from the Rig Veda to corroborate its word root analysis. Mantras in the main body of the Vedic texts were cognized in advanced states of meditative consciousness, and cognizers are credited with the mantras in the main body of the Vedas. Cognition proceeded from the Riks, through to the Atharva mantras, culminating in dialogues between cognizers and their students as captured in the Aranyaka and the Upanishads texts.

Six disciplines of learning arose to preserve the material from cognition for future generations. These are the Vedangas – the limbs of the main body of text. They are (a) Shiksha (b) Kalpa (c) Vyakarana (d) Nirukta (e) Chandas (f) Jyotish. Shiksha contains rules for phonetics and chanting and was emphasized in the past to guarantee accuracy and correctness in maintaining the Vedas through the oral tradition. Kalpa explains rituals and gives instructions on geometry and mathematics for the construction of fire altars. Computer scientists have been amazed with the precision and completeness of the grammatical rules of the Sanskrit language which are part of the Vyakarana limb. Nirukta explains the derivation of words from a basic set of verb roots. Knowledge of Nirukta rules and a mastery of its examples develops one's skills in understanding the nuances of selecting from multiple meanings for a Mantra. Chandas is the science of Vedic meters and explains the use of intonations. Jyotish covers three areas – Astronomy, Astrology and *Upaya*. Yaskacharya's Nirukta is an extant text on one of the six disciplines of learning while the Surya Siddhanta text is part of the Jyotish discipline of learning. During the past few centuries, the astrology and the upaya aspects of Jyotish have gained in popularity compared to the astronomy aspect. An earlier chapter was devoted to a discussion of the history of the Jyotish tradition.

Understanding the differences between the Vedic and modern scientific approaches helps us better appreciate the analysis presented in this chapter. The modern scientific approach distances an observer from his observations, while the Vedic approach gives equal importance to the process of observation, the observed, and the observer in understanding reality. The religious persecution of medieval Europe precipitated the separation of science from spirituality during a crucial phase in the development of European sciences, setting the tone for the modern scientific approach, but the line between science and spirituality was always blurred in older civilizations. Modern cosmology and quantum physics struggle to explain many phenomenon without the freedom to acknowledge consciousness as a possible scientific entity. Colonial translators too, were left with no option but to propose archaic ideas about the development of Vedic thought process. Scientists studying Vedantic texts today instantaneously recognize the similarities with modern quantum theories.

In order to understand the Vedic view of the cosmos, it is important to review the observer centric paradigm. If aspects of existence within the purview of human sensory experiences are considered part of the Bhu domain, indirect understandings are part of the Antariksha domain, and instinctively experienced subtleties are part of the Dyu domain. Vedic text frequently refers to the Bhu domain by the term *Idam*", meaning "this". Experts use special equipment to develop new concepts and verify them; astronomers conduct complicated experiments to prove that the shape of our galaxy is spiral. But the common man relies on the statements of experts to further his understanding of the universe, and fundamentally, these understandings are examples of the Antariksha domain of knowledge.

Even modern scientists resort to random selection or probability as the reason for certain phenomenon. In cosmology, a prominent example of this is the cause for the Big Bang. The Vedic approach classified these as cosmic impulses that cause energy to move in a specific way in the third domain. These impulses, according to the Vedas, lie beyond measurement. The existence of such impulses may only be inferred by the repeated manifestation of an associated experience. Human emotions and intuitive understanding, as well as the Big Bang are all attributed to such impulses. The Vedic seers classified these impulses as Devatas.

Learning and understanding happen in unique ways in the three domains. The domain of *Bhu* consists of experiences involving matter in its different forms. Comprehension in the domain of *Antariksha* is made possible through thoughts and concepts. Human language is the vehicle for the acquisition of this knowledge with *vac*, or speech, serving as the bridge between human brains. The nervous system plays a crucial role in all human understanding and experiences. Molecules comprise it, but awareness is made possible because of electrical currents passing over them. Knowledge about a remote corner of the universe is gained only by observing the electromagnetic radiation emanating from the location. The unseen world, both inside and out there, is perceived and understood only through energy. Cognition, then, is the process through which awareness glimpses impulses that cause the movement of energy. Cognitive knowledge is the domain of *Dyu* and is a part of cosmic consciousness.

A few simple examples of how Devatas are named follow here. The Devata Agni's role is implied wherever energy becomes manifest. Digestion is attributed to the digestive energy, an aspect of the Devata *Agni* as *Vaisvanara.*

Knowing is attributed to neurological energy, another aspect of Agni called *Jatavedas*. Light and heat which sustain life on the planet are the radiant energies, another aspect of Agni named *Surya*. *Vayu* may be understood as an aspect of Agni that drives meteorological phenomenon on Earth. The Vedic civilization venerated different aspects of energy in the cosmos by building a flame in a fire altar, creating a representative form of the otherwise abstract Devata Agni. The classification of Devatas is just a convention as can be seen from the fact that some mantras praise Agni to be the primary Devata while some others do the same for Vayu or *Indra* and so on.

However, the functional assignment to a Devata is indicated by a particular name. For example, within a complex system, the control function is attributed to Devata Indra. The word *indriya* denoting the human nervous system implies Indra's position to be in the human brain, the central control mechanism in the body. Conclusions about Indra's and Agni's position in the galaxy can be reached by using this example. Energy is omnipsent in the galaxy but is more intense in stars and thus represent the Devata *Surya*. The most intense group of stars highlights the fiercest aspect of energy and this substantiates the assignment of Agni to the Krittika region (containing Pleiades) of the sky. Indra's position must be near the galactic center, which scientists understand to exercise control over the level of star formation activity within the Milky Way galaxy.

The remainder of this chapter contains an analysis of the twenty seven Nakshatras. The analysis for each Nakshtra has three subsections: the first subsection presents the analysis of the names of the associated Devatas, introduces the main points from the Nirukta, and contains additional Vedic mantras that Yaskacharya used for clarity. It also contains suggestions for alternate English meanings to the ones employed by Lakshman Swarup (Swarup). Notes in the footer section explain the choice of alternate interpretations of selective words. The second subsection quotes the Nakshatra mantras with Paul-Emile Dumont's English translations (Dumont, 1954). This subsection includes Dumont's translations along with suggested modifications from an astronomical context. The third subsection reviews specific qualities of Devatas which can help identify astronomical phenomena resembling the characteristics of the Nakshatras. The identified astronomical phenomena are elaborated upon in the following chapter to help readers consolidate understanding of resemblances.

The first five Devatas in this chapter are aspects of a conducive environment for stellar or galactic formation. The star formation process consists of multiple phases, each of which is influenced by the next seven Devatas analyzed. The influences of the next ten Devatas dominate in particular structures of the Milky Way Galaxy. The subtlest aspects of the creation process of the cosmos are mirrored by the qualities attributed to the next two Devatas. The last three form a miscellaneous group.

Creative Energies

agniḥ

Nirukta introduces Agni to be the foremost leader, invited first during sacrificial rituals. Any object in proximity to fire becomes fire. The Rishi Sthaulāshīvi's opinion is 'It does not moisten and is therefore a drying agent". Rishi šākapūṇi on the other hand analyzes each phoneme of the word Agni to be from the following roots: to go, to shine or to burn, and to lead. The letter 'a' is from the root 'i' (to go), the letter 'g' from the root 'anj' (to shine), or 'dah' (to burn), with the root 'ni '(to lead). The following are the most relevant mantrās to Agni's role in the Nirukta.

agniḥ pūrvebhirṛšibhirīdyo nūtanairuta.
aa devāgum yeha vakṣati .

Rishis of ancient times sought out Agni. Modern folk (with quest) should do the same. (As) He is capable of drawing other Devatas here. Elaborating the second half of this mantra the Nirukta explains

sa na manyetāyamevāgniriti.apyete uttare jyotiṣī agnī ucyate.

This right here is Agni. Also over there the luminaries (in the sky) are Agni only. Another Mantra is even more explicit. It states that all is Agni only. Agni is variously called Indra, Mitra, Varuna, Suparna, Yama and Matarisva

indram mitram varunamagnimāhuratho divyah sa suparno garutman.
yekam sadviprā bahud vadantyagnim yamam mātariṣvānamāhuḥ.

Agni, commonly seen as a flame, is a proxy for the universal energy principle. In a quantum mechanical context, the entire creation, including what appears solid, is made up of a wave function (energy). Krittika provides the most intense experience of the power of Agni. Rishis may have listed this Nakshtra to be the first in the sequence due to the preeminence of its Devata Agni in the Rig Veda. The following is the mantra for Krittika Nakshatra:

agnirna pātu kṛttikā. nakṣatram devam indriyam[21].
idamāsām vicakṣaṇam. havirāsam juhotana.
yasya bhānti raṣmayo yasya ketavaḥ.
sa kṛttikābhirabhi samvasānaḥ.
agnir no deva suvite dadhātu

Let Agni protect us, (let) the Krttikas, the Nakshatra (of Agni), the divine manly power (protect us). Pour (O priests) this radiant oblation into their mouth. Let him, whose rays, whose banners, shine, to whom all these beings, all, belong, who wraps himself in the Krttikas,-let god Agni establish us in welfare.

This mantra treats Krittika and Agni as synonymous. It emphasizes the brilliant rays as well as the smoky coverings in the Krittika region. The word Krittika can be derived from the root kṛ – "to cut". Kṛttikā refers to a cutting/parting impulse. Alternate meanings of Krittika, according to the analysis in the Nirukta, are fame and food.

The most intense stars whose rays appear to be shrouded in smoke can qualify to be astronomical representations of Krittika and Agni. Pleiades, an open cluster of stars, fits this combination. The stars in Pleiades have an intensity which is seen in only a fraction of stars in our galaxy. The next chapter elaborates the characteristics of an open cluster where the gravitational forces are not as strong. Stars from it move away as if something is "cutting" (a quality of Krittika) the cluster apart. Of all open clusters, the Pleiades group is among the closest to the earth. Scientists believe that our own Sun may have started life in an open cluster and migrated to its present position. The reference to Krittika being a part of Indra is supported by the fact that all regions receive sustenance from the center, the region of Indra or the Jyeshta Nakshatra.

prajāpatiḥ

The word prajāpati is made up of two words, namely, praja - "subsequently born[22]"- and the word pati - "protector" or "supporter." The following Mantra in the Nirukta elaborates:

[21] Indriyam can be seen as a part of Indra. Therefore Krittika serves as a part of Indra
[22]Commonly translated as creatures

prajāpate na tvadetānyanyo viṣvā jātāni pari tā babhūva.
yatkāmāste juhumastanno astu vayam syāma patayo rayāṇam-

Prajapati, you enclose all created things. With whatever desire we sacrifice to thee, let that be ours. May we be lords of treasures!

Some translators consider the words "Brihaspati," "Brahma," and "Prajāpati" to refer to the same Devata. However this idea is not supported by the fact that each of them is assigned a different Nakshatra. The word Prajāpati, refers to an enclosure impulse that is evolving towards creation from within and being able to encompass and sustain that future creation. This interpretation lets us identify the astronomical phenomenon in the Nakshatra Rohini. The word Rohini may be derived from the root "ruh" – to develop/grow. Prajāpati is its progenitor.

The mantra for Rohini Nakshtra is.

prajāpate rohiṇī vetu patnī viṣvarūpā bṛhatī citrabhānu
yatha jīvema ṣaradassavīrāh̄ rohiṇī devyudagātpurastāt
viṣvā rūpāṇi pratimodamānā
prajāpatigum haviṣā vardhayantī.
priyā devānā-mupayātu yan̆am

Let Prajapati's wife, Rohini, the many-colored one, the great one, of bright appearance, enjoy (this oblation). Let her establish us in the welfare of the sacrifice, that we may live (many) autumns, with manly sons. The divine Rohini has come in the east, greeting with joy all appearances. Gladdening Prajapati with the oblation, let her, who is dear to the gods, come to the sacrifice

A better understanding is gained with a slightly different translation of the above. Rohini, the spouse of Prajapati, can be seen to be a reference to a world which is a large, conspicuous and luminous entity.

Let her (Rohini) provide an access to the Yagna by which the worlds survive for long. Devi Rohini, rose up earlier. Shining light on all the forms in the world. (She) with *havish*[23] causes Prajapati to grow. Let (she who is) pleasant to Devatas arrive in this Yagna.

Being a spouse, she (Rohini) supports Prajapati's impulse of moving towards a new creation. Creation is continuous and cyclical according to the Vedic

---

[23] An offering made in the sacrificial fire

tradition. Regions of the universe go through creation, growth, termination and finally towards another cycle starting with creation. Therefore we look for a growing astronomical phenomenon leading into another phase of creation. Red giant stars continue to expand until terminating their lives in a supernovae, spreading raw material for subsequent generations of stars. A red giant is a good representation of the Nakshatra Rohini, and aptly, we find one of these in this Nakshatra region of the sky. Aldebaran is among the the closest red super giants in the night sky, dominating other stars in this Nakshatra region. Its physical attributes match the epithets Brihati, large, and Chitra Bhanu, magnificent sun. The next chapter describes the process through which a red super giant expands into an enormous size before exploding into a supernova. After the raw materials from a supernova have mixed back into the interstellar medium, the support of *Soma*, the next Devata, is necessary to make the environment conducive for creation.

Somaḥ

The word Soma is derived from the root 'su' - to press; it denotes an effort to enrich by collecting diluted and scattered essence. Soma is the name of a plant from which an exhilarating juice of the same name is extracted. The ninth mandala of the Rig Veda is devoted to Soma, the Devata. He is the *Pavamana* -- "the purifier." Celestial Soma is the essence of creation, and its scattered nature is comparable to the sap in a plant. The sap becomes concentrated by juicing; strength is returned to creation by consolidating the scattered life force. The gathering of tiny drops of Soma from the Soma plant, and its offering in Yagna imitates the creative environment of the universe. Mantras praise Soma drops as bringing strength to the Devatas.

Soma complements the drying nature of Agni. Agni's impulse drives a man into action which eventually tires him. He then recoups his enthusiasm and gathers back his vitality (Soma's impulse) by relaxing. Indu is the holder of the drops of Soma. Yoga literature often compares the Agni-Soma pair to the Sun-Moon pair. The common translation of the word Mrigashira is "the head of a spotted deer." The word Mriga has the meaning "whose secondary markers are insignificant" and the word Shira has the corresponding meaning of "the most significant part of an entity." Thus the word Mrigashira denotes a region or a phenomenon containing a significant entity lying among the commonplace. The word Mriga derived from the root 'marg' meaning "to search," provides another clue to the Nakshatra being a starting point of a search.

The Nakshatra mantra for Mrigashira is

somo rājā mŗgşīršena āgann. şivan nakşatram priyamasya dhāma
āpyāyamāno bahudhā janeşu retaħ prajām yajamāne dadhātu.
yatte nakşatra mŗgazīršamasti.
priyagm rājan priyatamam priyāņām.
tasmai te Soma havišā vidhema.
şanna edhi dvipade şam catušpade

King Soma has come with Mrigashira (his Naksatra). The auspicious Nakshatra is his beloved dwelling place. Becoming strong in many ways among the creatures, let him give seed and offspring to the Sacrificer. To that one which is thy Nakshatra, Mrigashira, the dear one, O king, the dearest of the dear ones, to that (Naksatra) of thee, O Soma, we would offer worship with oblation. Be beneficent to our biped (i.e. our men), beneficent to our quadruped (i.e. our animals).

The phrases "giver of *retah*-vitality," "pleasant," "benevolent" are epithets of Soma in this Mantra. They highlight the rejuvenating quality of Soma.

Soma enhances the process of creation by taking it through a phase of recuperation. Soma impulses therefore should be recognizable in an environment becoming conducive to creation. Cold interstellar space, abounding with molecular clouds, are candidates to symbolize the Soma impulse because the creation of stars begin in them. The Orion region of the sky is one such region which is rich in raw materials where stars can form. It abounds in "shapeless" molecular clouds. The coldness provides the right conditions for molecules to stick together. When a shock wave moves in from a neighboring area these molecules can move even closer. The extraction of the Soma juice is an apt imitation of shockwaves nudging the molecules together. The coolness of the container in which the Soma juice is collected is analogous to the cold environment existing all over the Orion region of the sky. The next chapter explains how the collapse of a cloud is the first step in the formation of a star. In this Nakshatra, scientists have identified many cold molecular cloud regions where a collapse is most likely to begin. Mrigashira is a fitting name for a Nakshatra hosting the closest zones from the earth in space with cold expanses of such molecular clouds. Rudra, the next Devata, causes turbulence that rolls through cloud regions such as Orion to catalyze a collapse.

rudraħ

Rudra is closely connected with Agni, the god of fire. Both are powerful and resplendent, with great capacities for destruction. Rudra is so called because he bellows (rauti), or because he runs (dravati) vociferating, or he roars - root rud. Quoting from the *Kathaka* and the *Haridravika* texts, Nirukta expresses the idea that roaring is a characteristic of Rudra. Rudra is often praised with Soma and Pusan. Rudra is also associated with Maruts. The root Ru is to sound, to vibrate. The word Rudra means to shed, to flow, to inflict as can be seen in the following mantra

yā te didyudavasŗšţā[24] divaspari kşmayā carati pari sā vŗņaktu nah.
sahasram te svapivāta[25] tanayešu[26] rīripah. ţ

> May that cutting weapon of yours, which, created from regions beyond, moving together with earth, avoid us. Your trustworthy speech is like a thousand medicines. Hurt not our sons and descendants.

Turbulence in air currents through the vocal chords creates speech. They instead cause destruction when storming through nature. The Nakshatra mantra for Ardra praises the qualities it shares with its Devata Rudra:

ārdrayā rudra prathamāna eti. şrešto devānām pati agniyānām.
nakştramasya havišā vidhema.Mā naħ prajāgum rīrišanmota vīrān,
hetī rudrasya pariņo vŗņaktu.
pramunchamānau durithāni vishvā.
Apāghaşagumsannudathāmarāthim

Along with Ardra (his Naksatra), Rudra, spreading himself, goes, the best of the gods, the lord of cows. To his Naksatra we would offer worship with oblation. May he not harm our offspring, may he not harm our men. Let the dart of Rudra pass us by. Let Ardra, the Naksatra (of Rudra), enjoy our oblation. Shaking off all evils, let them both (Rudra and Ardra) drive away the wicked, the enemy.

With the epithet şrešto, Rudra is addressed as the best among the Devatas and foremost among energies. The phrase "pramunchamānau durithāni vishvā" can easily be translated as hurling out turbulence in the universe. We find a

---

[24]*didyut* derived either from do – to cut, or from dyu – to assail or from *dyut* - to shine

[25]*svApta vacanam* – abundant speech *bheSajA mA nastokeSu. Tok*a is derived from tud – to push

[26]derived from *tan* - to spread

discussion on the turbulence of Rudra in the Nirukta. This Nakshatra mantra seeks Rudra's help in driving away the enemies. Nirukta describes Rudra's healing speech to be his softer side. Speech arises from turbulence moving through the vocal chords. It is much softer and kinder and is an example of Rudra's turbulence which is benevolent. Another common translation of the word Ardra is moist. The word Ardra shares the root dra - to run ormove, with the word Rudra. Because of this common root, Ardra can also mean a flow, a movement.

To represent Rudra's impulse in astronomy we look for turbulence in space that is benevolent to the cosmos. Astronomers using ESA's Herschel space observatory have found dense filaments of gas all around in space; each filament is about 20000 times the distance of Earth from the Sun. Explaining the consistency of the widths, through computer models, astronomers theorize that sonic booms traveling through clouds lose energy and leave filaments of compressed material. Stars form in these filaments like beads on a string. One cause of sonic waves in the vicinity of our Sun is the star Betelgeuse which is hurling through space at a fast pace. The vicinity of the star Betelgeuse has a fiery nature, as if over powered by Rudra. Complex convection currents within the huge shell of the star create giant outbursts of plasma spreading out over six times the diameter of the star. Scientists have detected water vapor in this atmosphere. This corroborates the meaning of moisture for the word Ardra. The almost vacuum like object, only way too hot, lying in Ardra Nakshatra is a complete contrast to the densest object in our Galaxy, a black hole in the diametrically opposite Nakshatra Mula.

aditī

Aditi is composed of the root Da – "to be bounded" and the negation prefix "a". The word Aditi is to be understood as "unbounded." Aditi is heaven, Aditi is atmosphere, and Aditi is mother, father, and son. Aditi is all the Devatas. Aditi is what is born and what shall be born. Nirukta describes the relationship of Aditi and the Adityas whose characteristics it elaborates. In one famous reference, Aditi attends to the four adityas -- Daksha, Mitra, Varuna and Aryama. The Adityas inherit the characteristics of their mother Aditi. Here is a contradiction related to Aditi. Daksha was born from Aditi, and Aditi sprang into life from Daksha. "How can this be possible?" Nirukta raises this question and looks for an explanation. Subsequent Vedic texts refer to Daksha as a

Prajapati. A paragraph below provides an astronomical explanation for the contradiction.

The Nakshatra mantra for Punarvasu starts with the phrase "once again":

punarno devyaditisspŗņotu. punarvasūnaȟ punaretām yaňam.
punarno devā abhiyantu sarve. punaȟ punarvo havišā yajāmaȟ.
evā na devyaditiranarvā. vişwasya bhartrī jagataȟ pratišţā.
punar no devaȟ abhiyantu sarve.
devānām apyetu pāthaȟ.

Again (punar) let the goddess Aditi deliver us (from evil); again (punar) let the two Punarvasus (who are her Naksatra) come to our sacrifice. Again let all the gods come to us. Again and again we honor you (O gods) with oblation. Like a swift mare let the goddess Aditi, the irresistible one, the supporter of the universe, the foundation of the world, gladdening the two Punarvasus with the oblation come to the place that is dear to the gods.

This mantra highlights the phrase *Punaȟ Punaȟ* - "again and again." The name of the Nakshatra contains the word *punar* meaning "again" and the word *vasu* meaning a residence of life force. Aditi is the universal progenitress and the foundation of the world. Punarvasu clears the path (pāthaȟ) for the arrival and manifestation of Devatas.

We look for an astronomical phenomenon supporting the "Aditi contradiction" discussed above. We find this in the relationship of a planetary nebula to its parent star and the subsequent generations of stars. The identified astronomical region also needs to exhibit Aditi's nature of a progenitress. Daksha and the other Adityas -- being born of Aditi -- appear to refer to different stellar phenomena evolving from a common source. An understanding of the process of star formation, suggests different "Adityas" may just be references to the phases in the formative process. Devatas, namely Aryama, Bhaga and Varuna, Mitra and Pusha, are considered to be Adityas and have been assigned as the Devatas of a few Nakshatras. A huge planetary nebula, Abel 21, four light years across, in the sky region of the Nakshtra Punarvasu is a token of Aditi's role in incubating the future of the cosmos. Brihaspati, the Devata of the next Nakshtra leads the evolution from an embryonic state to one of maturity

bṛhaspatiḣ

Brihaspati is the supporter and the protector of the vastness principle. Mandalas four, seven and ten of the Rig Veda contain hymns to Brihaspati. He is the friend of Indra and the enemy of Vala. Vala and Vritra are the negative principles indicating hoarding that block the creation process. They have been customarily personified as the clouds that horde water. In the context of meteorology, Indra and Brihaspati counter these two negative forces to release the waters. Vritra and Vala continue to grow unless challenged, as their nature is to lock up whatever creates abundance on the Earth and in the cosmos. A parallel in the deep sky for the Vala principle is any barrier to the growth of the visible universe that is annulled by the Brihaspati principle. Indra's role in maintaining the star formation process in the Milky Way, can be seen in the section on Indra, later in this chapter

The role of the Brihaspati principle is better understood from the Nakshatra mantra for Tishya (Pushya)

bṛhaspatiḣ prathamam jāyamānaḣ.
tišyam nakşatramabhi sambabhūva.
şršţo devānām pṛtanāsujišṇuḣ dişonu sarvā abhayanno astu.
tišya purasthāduta madhyato naḣ. bṛhaspatirna paripathi paschat.
bādhetāndvesho abhayam kṛṇutām. Suvīryasya patayassyāma

Brihaspati, when first born, went to Tisya, his Naksatra, he, the best of the gods, the one who is victorious in battles. Let it be security for us in all quarters. Let Tisya protect us in front, and also in the middle; let Brihaspati protect us in the rear. Let them both (Brihaspati and Tisya) drive away hostility. Let them give us security. May we become lords of a host of brave men!

Brihaspati here, is described as the supreme among the Devatas, triumphant in hostile encounters, providing protection from all directions He provides relief from the divisive forces *bādhetāndvesho* so that we become endowed with proper strength. Tishya is from the root "shta" - to stay. It means "bringing stability." The alternate name for this Nakshatra is "Pushya" which means "providing nourishment."

We look for an astronomical principle in this Nakshatra region of the sky that nurtures creation, protecting its state of nascence from external disturbances and helping it mature. Brihaspati's influence stands out in two phenomena, one relating to the formation of a star, and another relating to the formation of earth like planets. One of the most important first steps in the formation of a star is the hastening of an initial collapse. Scientists have gained a preliminary understanding about the role of magnetic forces in speeding up the collapse process. A collapse that doesn't collect enough material will result in brown dwarf stars that cannot emit light. Brown dwarfs and even smaller Jupiter sized stellar bodies are failed stars whose mass is insufficient to support nuclear fusion. The next chapter describes the astronomy of star formation. Large sized planets forming, beyond what astronomers term the frost line protect nascent terrestrial planets from being bombarded by material from space. In the language of the Vedas, Brihaspati is equated with the impulse that nourishes and protects the process of growth. This impulse exhibits itself during the stellar and the planetary formation processes. Cancri 55 is a complete solar system in the Tishya Nakshatra region where astronomers have detected four Jupiter like planets.

Planets form alongside the central star. The stabilization of planetary orbits is connected with the next Devata, Sarpah.

sarpāḣ

Vedic texts sometimes reference the names of animals to draw attention to prominent characteristics they share with physical phenomenona. Remembering this subtlety while interpreting the Vedic Mantras can provide more sensible translations. The word Sarpa is derived from the root Srp which has the multiple meanings – to glide, to move. The word Sarpa draws one's attention to the presence of a movement similar to that of a snake. Nirukta mentions the Devata Sarpa Raagyi to be associated with Surya, the sun. Translations of text, especially referring to Sarpa in the Sky turn out senseless when the movement trait is ignored. The word Sarpa sometimes refers to the quality of a laser sharp awareness in contrast to a broader awareness indicated by the word Garuda, the Eagle.

The word ashresha (ashlesha) can be derived from the root Srp. Ashresha is translated as "to cling" or "to intertwine". The word sarpa too is derived from this root. The Nakshatra mantra for ashresha is

idagum sarpebhyo havirastu juşţam. āşrešā yešāmanuyanti cetaḣ. ye antarikşam pŗthivīm kşiyanti. te nassarpāso havamāgamišţhaḣ. ye rocane sūryasyāpi sarpāḣ. ye rocane sūryasyāpi sarpāḣ. yešāmāşrešā anuyanti kāmam.

tebhyassarpebhyo madhumajjuhomi.

Let this oblation be agreeable to the Serpents, whose will the Ashresas (who are their Naksatra) obey. Let the Serpents who inhabit the atmosphere, (and) the earth, come quickly at our invitation. And to the Serpents who are in the luminous sphere of the sun, to those who follow the goddess Sky, to those whose desire the Ashresas (their Naksatra) obey, to those Serpents I offer a sweet oblation.

In the phrase *ye antarikṣam pṛthivīm kṣiyanti*, we observe the use of the root *kshi* – being secretly hidden or being unseen. The nature of a snake is to lie hidden in the ground. But a reference to this quality in space (between the earth and the sun) can only refer to an invisible force. On the other hand, in the domain of the luminaries (divi), a different aspect of Sarpa is mentioned. The use of the verb 'follow' here, means Sarpa has a smaller role.

Looking for representations of Sarpa in astronomy is relatively easy. Its presence is indicated wherever angular momentum is seen in circular motion, such as in the swirling molecular clouds and the orbits of the moons of planets. The principle of conservation of angular momentum is important in physics; centripetal force, sustaining any curvilinear motion, may be indicated in a Vedic passage using the word Sarpa. The slithering, winding movement of a snake is not possible without this centripetal force. The former is a reminder of the general principle of the latter. Subtleties like this one are often lost in word for word translations from Vedic texts. We find in the Nakshatra Ashresha a symbolic association with Sarpa in the Epsilon Hydrae group of stars whose mutual movements resemble that of a group of slithering creatures. This Nakshatra thus draws one's attention to a phase of creation where angular momentum plays a crucial role. This is seen in the formation of a stellar system when nebulous angular momentum transforms into the ensuing spin and orbital paths of the resultant planets.

A planet forms from a number of planetisimals, aggregating together, terminating their separate existences. The function of Pitru Devatas reminds one of this principle of nurtured evolution.

pitṛ

The word Pitru is a colloquial reference to ancestors whose consciousness has separated from material existence but not yet the subtle. People revere their ancestors as Pitrus Devatas. However, Vedic texts make a distinction between the Pitrus as Devatas and those as ancestors. The adjective *Sasvata*, meaning permanent is used to avoid this confusion. There are seven of the Sasvata Pitru Devatas- Anala, Soma, Yama, Aryama, AgniShvAtta, BarhiShada, and Somapa. Of these Yama and Aryama find more frequent mention in the Mantras related to the Pitrus. The Pitru Devatas facilitate the passage of consciousness from the domain of the seen to the unseen. Their influence can be inferred just as one infers the influence of Sarpa Devata in the cosmos, namely, understand an abstract through a more common phenomenon. The Nakshatra Magha is associated with Pitru Devatas. The standard dictionary meaning of Magha is gift or bounty. The word magha can also be derived from the root gha (to terminate) which with a negation prefix "ma" may emphasize a transitory state. Vedic texts names one of the sub energies of Yama, associated with ancestors, as *ChitraGupta* which can be translated as a "hidden record" or an invisible imprint.

The Nakshatra mantra for Magha highlights the "burnt out" aspect of existence

upahūtā pitro ye maghāsu. Manojavassukŗtassukrityāħ.
te no nakşatre havamāgamišţhaħ.
svadhābhir yaňam prayatam jušantām.
ye agnidagdhā ye'nagnidagdhāħ. ye'mumlokam pitaraħ, kşiyanti.
yāgşca vidmayāgm u ca na pravidma.
maghāsu yaňagm sukŗtam jušantām

Let the Fathers who are invoked beside the Maghas (their Naksatra), (the Fathers) who are as swift as thought, doing good, and performing their duties, let them, beside their Naksatra, quickly come at our invitation; let them freely enjoy the well-prepared sacrifice. Let the Fathers who are burnt by the fire, and those who are not burnt by the fire, those who inhabit yonder (heavenly) world, those whom we know and those we do not know, let them, beside the Maghas (their Naksatra), enjoy the well-prepared sacrifice

The verb *kşiyanti* indicates an existence that is hidden; the same concept is emphasized by the phrase "Of whom we know well and also not know." Mantras often use an analogy of something which is more familiar to clarify something abstract, as observed in the case of Sarpa Devatas. As seen in the previous paragraph the word Pitrus refers to two paradigms. The mantra's intent may be to infer about the nature of the Pitru Devata through one's understanding about the Pitrus, the ancestors.

Following such an analogy we look for an evocative astronomical entity which has stopped participating in active existence having steered the next generation to maturity. The phrases "burnt out," "exhausted their energies," and "out of here," found in the Nakshatra Mantra, fit well, the characteristics of a category of satellite galaxies which astronomers know to be devoid of star formation activity. Extinct satellite galaxies have "gifted away" their content to the Milky Way galaxy enabling it to grow to its current shape and size. Their essence continues to exist even though their independent existence has ended; the Leo I dwarf is a reminder of this enigma.

aryamā

The word Aryama is derived as aha and yama – "maintaining the day and night cycles." Mantras recognize him as a ruler of time. A mantra quoted in the Nirukta highlights Aryama as a solar entity with many chariots, whose movement is unimpeded, and whose path is slow and unobstructed by anything. Seven rays extract juices on his behalf. He officiates on the birth of diverse forms, activities[27] and sunrises. He shares some common traits with the six other Adityas listed in Rig Veda.

havišmāgm āvivāsati. aryamādityaḣ. bahurathaḣ. atūrtapanthāḣ atvaramāṇapanthāḣ. saptahotā saptā smairaşmayorasānabhisannāmayanti. višamarūpešu janmasu karmasūdayešu

The Nakshatra Mantra for Falguni (Purva) is presented below

gavām patiḣ falgunīnāmasi tvam. tadaryaman varunamitra cāru. tam tvā vayagm sanitāragm sanīnām. jīvā jīvantantamupa samvişema. yenema vişvā bhuvanāni samjitā. yasya devā anusamyantu cetaḣ. aryamā rāja'jarastu višmān. falgunīnāmŗšbho roravīti

Thou art the lord of cows, (the lord of) the Phalgunis (the red ones) (i.e. the First Phalgunis, who are thy Naksatra). That, O Aryaman, O Varuna, O Mitra, is dear (to thee). May we, alive, sit down near thee, the living one, and the bestower of gifts! He by whom all these beings have been conquered, whose will the gods obey, Aryaman,

[27]Because of his role as the witness of new activities, Aryama is invoked in wedding ceremonies.

the king, the undecaying one, the powerful one, the bull of the Phalgunis, bellows loudly.

The alternate meaning of *Gau*, the sun, makes better sense in this mantra rather than the commonly applied meaning of the cow. The word *viśmān* -- one who expands swiftly -- provides another clue about the astronomical phenomenon that is involved. The reference to the animal *rishabha* -- the bull, here -- can be a figurative reference to the impulse of supplying the seed for a new entity to develop. Aryama is an aspect of the Sun, becoming alive with a powerful roar, whose influence expands fast, is pleasing, and is similar to varuna and mitra.

We look for an astronomical phenomenon where planets may still be shrouded in debris waiting to be cleared by emissions from the central star. Unless the super charged emissions clear the path, sun rise and sun set may not be observable from the planets around a star. A new star, just having started fusion in its core, blows out strong plasma winds rich in charged particles that vaporize dust and debris remaining in the protoplanetary disc surrounding the star. This makes the young star visible from its own planets where days become distinct from nights. A source of strong plasma wind can be a symbol of Aryama's power and intensity. The Nakshatra Mantra for the Falguni Nakshatra also refers to the act of seeding which in this case can be interpreted as readying a planet for life. High energy stellar wind is similar to galactic wind blowing from the center of an active galaxy. Leo cluster in the Falguni Nakshatra region includes three quasars around its central galaxy. Quasars emit the most energetic of plasma winds seen in the universe. The next chapter describes the birth of a new star and the emissions from a young star in more detail.

bhagaḣ

Bhaga is derived from "Bha"and "ga" -- attaining growth or shine. Bhaga's time is precedent to the sunrise. The following famous mantra is addressed to Bhaga.

prātarjitam bhagamugragm huvema vayam putramaditer yo viddhattā

> May we invoke the early-conquering Bhaga, the fierce son of Aditi, and the on who supports all. The mantra continues with a mention of Yama, the lord of death, making a prayer to Bhaga. Bhaga is considered to be blind – not visible, while not risen. In another mantra addressed jointly to Aryama, Pusha and Bhaga the epithet *Karulati* – one without teeth is thought to apply to

Bhaga. The most frequently applied adjective for Bhaga is "Bestower". Bhaga along with Aryama, Mitra, Varuna, Pusha, Savita and Amsa are considered to be the seven stellar principles born out of Aditi.

The Nakshatra mantra for Bhaga begins with

şrešţho devānām bhagavo bhagāsi.
tatvā viduħ falgunīstasya vittat.
asmabhyam kşatramajaragm suvīryam. gomadaşvavadupasannudeha.
bhagoha dāta bhaga itpradāta. bhago devīħ falgunīrāviveşa.
bhagasyettam prasavam gamema. yatra devaissadhamadam madema

Thou art the best of the gods, O blessed Bhaga. So the PhalgunIs (i.e. the Second Phalgunis, who are thy Naksatra) know thee. Do thou know it? Bring us here undecaying ksatriya power, and a host of brave men, with cows, and with horses. Bhaga is the giver, Bhaga is the bestower; Bhaga has taken possession of the divine Phalgunis (who are his Naksatra). May we obtain the impulse of Bhaga, so that we may feast with the gods!

An alternate meaning for the word *kshatra* is a region. The words *gau* and *ashva* may refer to principles in cosmos, rather than the animals themselves. The Nakshatra mantra emphasizes the point that bhaga is the shining and growing entity, giving before, giving now, and being born to give again. The name of the Nakshatra Falguni conveys Bhaga's bounty. A possible derivation of the word Phalguni is from Phal and guNa -- a region reaching fruition again and again.

We look for an astronomical region that is bountiful, maybe teeming with newly forming stars. It is also a region with brilliance based on the meaning of the word Bhaga -- shining bright. The visibility aspect of the eerging radiations must also be a characteristic of the region. All these are found in the Whirlpool galaxy abounding in star burst activity. Thousands of newly formed stars lie covered within HII regions, and they gradually become visible as the HII cover clears out. However, the HII regions themselves are some of the brightest objects in the universe. The intense star burst activity in the Whirlpool galaxy region produces a rich supply of elements essential for the formation of habitable planets.

savitṛ

Savitar is the Devata who stimulates the human intellect, clearing the darkness of doubt. The famous Gayatri mantra praises the stimulating impulse of

Devata Savitar. The Nirukta quotes the following mantra identifying Savitar's role in the domain of the atmosphere

savitā yantraiḥ prithivīmaramṇaskambhane savitā dyamadrumhat...

> Savitr has fixed the earth with supports; Savitr has fastened heaven in unsupported space; Savitr has milked the atmosphere, shaking itself like a horse, and the ocean bound in illimitable space.
>
> Nirukta also highlights a Mantra from the Rig Veda
>
> viṣvā rūpāṇi pratimuncate kaviḥ prāsavid bhadram dvipade catuṣpade vi nakamakhyatsavitā vareṇyo'nu pranamušaso vi rājati
>
> The wise one puts on all forms. He has generated bliss for the biped and the quadruped. Noble Savitar has looked on heaven. He shines bright after the departure of dawn. His time is that when the sky, with its darkness dispelled, is overspread by the rays of the sun.

The Nakshatra Mantra for Hasta invites Savitar to emerge

āyātu devassavitopayātu. hiraṇyayena suvṛtā rathena.
vahan, hastagm subhagm vidmanāpasam.
prayacchantam papurim puṇyamaccha.
hasta prayaccha tvamṛtam vasīyaḥ. dakṣiṇena pratigṛbhṇīma enat.
dātā-ramadya savitā videya. yo no hastāya prasuvāti yaňam.

Let the god come, let Savitar approach with his golden, well-rolling, chariot, carrying hither Hasta (his Naksatra), the lovely one, who works skilfully, who liberally gives, and is holy. Let Hasta give us immortal, excellent, wealth. We take it with the right hand. May I reach today the giver Savitar, who will further our sacrifice for the benefit of Hasta (his Nakshatra).

The meaning of the word *ṛtam* is orderliness. The word *dakṣiṇena* means from the right side[28] . This Mantra refers to particular hand movements, "*prayaccha*" -- giving and "*pratigrbhnima*" -- taking. The common iconic representation of *hasta* is a closed fist. The act of giving is done with an open fist, fully revealing the gift being held in the palm. This mantra invokes the giving aspect of Savitar. His bounty is in the vicinity but yet, not visible. He graces something

[28]*Dakshin*a also refers to one kind of gift

into visibility. A similar sentiment is seen in the famous Gayatri mantra addressed to Savitar.

Something is shining but remains veiled. The veil is cleared by the brilliance of Savitar. We look for something that is veiled in the Nakshatra Hasta. This Nakshatra too, may refer to a phenomenon in the stellar formation process, namely, when a newly formed star starts shining bright but needs a subtle transformation before the brightness is experienced. Is this subtle transformation the evolution of life from matter? Only life can experience light. Savitar is associated with discerning intellect. An approximate astronomical object exhibiting the characteristics of this Nakshatra could be NGC 4552, which belongs to a class of AGNs that are veiled by an opaque shell. Scientists estimate 20% AGNS in the universe to be in this group. The next devata refines and chisels the amount of brightness available after a bright star is formed.

tvaštā

The word Chitra translates to "conspicuous" or "manifold." Tvashta may be derived from the root tvist - to shine, or from tvaks - to do. The following mantra is addressed to Tvashta:

> ye ime dyāvāpŗthvī janitrī rūpairpişdabhuvanāni vişvā

> Who adorned the heaven and earth, the two progenitors and all the worlds with forms? Some feel that Tvashta is an atmospheric Devata. Rishi *şākapūni* says that Tvashta is Agni and quotes the following Mantra addressed to him

> āvişţyo vardhate cārurāsu jihmānāmūrdhvaħ svayaşā upasthe.
> ubhe tvašţurbibhyaturyamānāpratīcī simham prati jošayete.

> The diffuser of light, the beautiful one grows among them, elevated by his own glory in the lap of the oblique[29]. With his glory, he attends high above the oblique/dim. Both being afraid of Tvashta, whose energy is beginning to

[29]*Jihmam* is derived from ha (to be bound) – alternate translation makes a lot more sense. Refer to Nirukta for the original translation.

be activated, reversing[30] their trend they begin aligning[31] with the force as strong as a lion. They two here are references to sky /earth, day/night, two wooden sticks to make fire.

The Nakshatra mantra for Chitra is

tvaštā nakṣatramabhyeti citrām.
subhagm sasamyuvatigm rocamānam. Niveṣayannamṛtānmartyāgṣca.
rūpāṇi pigmṣan bhuvanāni viṣva. tannastvaštā tadu citrā vicaštām.
tannakṣatram bhūridā astu mahyam.
tannaḥ prajām vīravatīgm sanotu.
gobhirno aṣvaissamanaktu yañam.

Tvashta unites with Citra, his Nakshatra the splendid young woman with beautiful hips, he (Tvashta) who brings to rest immortals and mortals, and fashions all beings into shapes. So let Tvashta, and so let Citra (his Naksatra), look at us. So let this Naksatra be liberal to me. So let it procure us a progeny abounding in manly sons. Let it beautify our sacrifice with cows and horses.

The phrase *rūpāṇi pigmṣan* refers to the act of chiseling to create distinct forms. It is somewhat similar to the Nirukta statements about Tvashta as one who diffuses light. In the Nirukta descriptions, Tvashta tones down excess brightness and causes other forces to become aligned with his purpose.

Chitra is "picture perfect." Tvashta instills beauty in the creation and creates definite forms from what is otherwise vague. A story in the Rig Veda relates to the marriage of Tvashta's daughter to Vivasvan, the Sun. A Brahmana text elaborates on this story and describes how Tvashta intervened and chopped out the excess brightness from Vivasvan so that his daughter could live comfortably with Vivasvan. This story presents the picture of tvashta as someone making objects a bit more pleasant to view by trimming out excess brightness, chiseling away excess mass. Intriguing phrases are found in the Mantras in the Nirukta describing Tvashta. They are easy to understand based on the concept of heaven (Dyu) and earth (Bhu) as introduced in an earlier chapter.

---

[30]Pratici is translated alternately as turning back, the opposite of Arvaci. This is also based on the Nirukta explanation *Pratyakte simham sahanam pratyAsevete*- Sahanam can be translated as "able to withstand." Simham can be translated as the vanquisher.

[31]Alternate meaning for *pratyAsevete* - attend upon

Looking for an astronomical phenomenon equivalent to Tvashta's pruning effect on a source of light, we find it in the phenomenon of polar jets that provide the finishing touches to the stellar formation process. They assist in the removal of an accretion disk from the immediate vicinity of a new star. This removal prevents the buildup of excess angular momentum in the new star. This sounds very similar to the function of Tvashta. Spectacular polar jets are seen when the material from the accretion disk get ejected. The next chapter discusses the formation of a polar jet both around a star and also around the center of an active galaxy.

Structural Energies

vāyuḥ

Yaskacharya, drawing from a prevalent collection of ideas, categorized Devathas into three main classes in his Niruktha. This classification scheme recognizes the influence of Devatas in the phenomena in three domains, namely, a) earth bound b) atmosphere bound and c) sky bound. This classification provided a simple way to interpret Vedic text in the context of popular ideas about the physical world during Yaskacharya's time. According to this classification, there are only three Devatas. All the other Devata names represent the diverse functions of these three, namely, Agni, Vayu/Indra, Sun. The analogy is to a person being called a mayor, a clerk, or an officer based on one's role in society. Rishis assigned names according to the function of a Devata.

In a three domain classification system, Vayu is the controller of all meteorological phenomenon. In his sphere, the other Devatas become his appellations. The layperson of older times could appreciate Devatas as forces of nature -- winds, clouds, thunder, lightning, rain. However, limiting the interpretation of Devata names to representations of the forces of nature on Earth leads to absurd translations of many Mantras. Nakshatra mantras relate to phenomena in space. References to Devatas in these mantras must adhere to the third class, namely, sky bound phenomena. Most translations in the past have ignored sky bound phenomenon, focusing on the meteorological domain. Specific characteristics of Vayu, associated with the Nakshatra Nishti, may be more appropriate in an astronomical sense than the generalizations commonly attributed to him. Discoveries in modern astronomy provide us an opportunity to update earlier translations related to Devatas from this angle.

Vayu is derived from va - to blow, or from vi - to move. According to Rishi *Sthaulāṣṭhīvi* it is derived from the verb i - to go, the sound v being insignificant. The following stanza is addressed to him.

Vāyavā yāhi darşateme somā arnkŗtaħ. Tesām pāhi şudhī havam

Come, O beautiful[32] Vayu, these Soma juices are ready. Drink them, hear (our) call. Some translators have the opinion that atmospheric nature of Vayu is presented in this Mantra. The following is another Mantra:

āsarāņāsaħ şvasānamacchendram sucakre rathyāso aşvaħ.
abhi şrava ŗjyanto vehayurnū cinnu vāyoramŗtam vi dasyet.[33]

May the linear swift movement created by Vayu add up and become a massive energy (of Indra) which is needed to transform untapped resources[34], gradually assembled over time.

Sayana, a famous commentator of the 12th century, interprets the word Nishtigra to refer to Aditi, implying that Nishti is diti (translated as bounded). The word nishtan refers to a thunder/roar. From this point Nishti appears to relate to something that is thundering and roaring. The Nakshatra Mantra for Nishti begins with describing the association between Vayu and Nishti

Vāyurnakşatramabhyeti nišţyām. tigmaşŗmgo vŗšbho roruvāņaħ.
samīrayan bhuvanā mātarişvā. apa dvešāgmsi nudatāmarātīħ. tanno vāyustadu nišţyā şŗņotu.
tannakşatram bhūridā astu mahyam.
tanno devāso anujānantu kāmam. yathā tarema duritāni vişvā.

Vayu unites with Nistya, his Naksatra, he, the sharp-horned bull, loudly bellowing. Shaking the worlds, let Matarishvan[35] (i.e. Vayu) drive away hatreds and hostilities. So

[32]Worthy of being seen

[33]Refer to Nirukta for the original translation. In the translations here, the words Ratha and Ashva are figurative and convey the influence of a Devata. Influence can be precieved through a result and a carrier in the physical domain. Ratha is a carriage of the impulse and Ashva's movement represents the realization of the impulse.

[34]The word anna here is a reference to an environmental supply of nourishment and the phrase "*navam ca puranam ca*" is considered to reflect something that is collected over a period of time

[35] This epithet *Matarisvan* also applied to Agni as a carrier

let Vayu, and so let Nishtya (his Naksatra), hear us. So let this Naksatra be liberal to me. So let the gods grant us the object of our desire, that we may overcome all miseries

Analyzing a pair of key phrases allows us to grasp the astronomical significances in this Mantra. The phrase *tigmaṣṛmgo vṛšbho roruvāṇah* refers to the intensity of a bull – bellowing/roaring. The phrase *apa dvešāgmsi nudatāmarātīh* refers to the act of challenging divisive forces. The phrase *yathā tarema duritāni viṣvā* can be understood better with the translation "by which the bad course of development can be avoided".

Strong plasma winds blow from the center of the Milky Way galaxy. This wind is not only forceful but its plasma literally tears through the clouds along the way. Scientists have put forth the idea that black matter is simply molecular/atom hydrogen which has no spectral signature. Vayu as *Matarisvan* carries enormous energy. Blowing plasma winds spilt molecules, like a bull piercing through objects in its path. Galaxies with weak plasma in their galactic winds have less intense star formation activity. Galactic winds blowing from the center of our galaxy contain strong plasma which arrive at the heliosheath of our Sun. The heliosheath blocks the charged particles in the plasma winds. A gentler wind containing uncharged particles -- atoms and molecules -- percolates through. The next chapter describes how this wind is deflected and appears to arrive from the direction of Libra where the Nishti/Swati Nakshatra is located.

indrāgnī

Nirukta does not analyze the name Indragni. Paired Devata names are common in Vedic text, for example, mitra-varuna and indra-agni. Indragni is a grammatical compound and thus may refer to the common roles of two Devatas. However this paired name in the Vedic texts represents a distinct Devata. Devatas with paired names often have their own Rishis and mantras to be used in rituals associated to them; for example, the rituals associated with Indragni do not use the mantras for Indra or Agni. Mandala 3 of the Rig Veda contains a Mantra in the 13th stanza, addressed to Indragni

indrāgnī rocanā divaħ pari vāješu bhūšathaħ vtad vāmceti pra vīryam

Pleasant and in the transcendental domain of horses known for spirit, speed, vigor (let) hundreds of world imbue that energy. This mantra says that

> Indragni matches the capacity of Agni -of creating forms- and that of Indra- to mobilize from within.

The word Vishakha relates to branching. Some translate the word to be "without branches" and some translate the word to "branched." The Vishaka Nakshatra Mantra starts with

dūramasacchatravo yantu bhītāh̆. tadindrāgnī kr̦ņutām tadvişākhe.
tanno devā anumadantu yaňam. paşcāt purastādabhayanno astu.
nakşatrāņāmadhipatnī vişākhe şrešţhāvindrāgnī bhuvanasya gopau.
višūcaşşatrūnapabādhanānau. Apakşudannudatāmarātim.

Let our enemies, terrified, flee far away from us. Let Indra and Agni, let the two Vishakas (who are their Naksatra), accomplish that (for us). So let the gods rejoice over our sacrifice. Let security be ours in the rear and in front. The two Vishakhas are the sovereign queens of the Nakshatras; the most excellent Indra and Agni are the protectors of the world. Driving away the enemies in all directions, let them expel hunger, the foe.

Alternate interpretations of two phrases of the Mantra provide a good clue to the astronomical context -- *višūcaşşatrūnapabādhanānau* -- as dispersing negative forces and *apakşudannudatāmarātim* as starving and driving away the opposition. Indragni in Vishaka Nakshatra drains out opposition, overpowers and dissipates them while protecting positive forces.

The far side arms of our galaxy join the far end of our galaxy's central bar in the Vishaka region. This junction is a conduit for ionized hydrogen pumped away from the galactic center into the arms of the Milky Way. Unused star material, rich in neutral hydrogen, is pushed back into the central region from the galactic arms via the junction. Scientists call the pumping and pushing motions the phenomenon of density waves. Without these density waves, star material within a galaxy can stagnate. Vishakha is one of the two junction points for density waves to influence activities in the spiral arms of the Milky Way galaxy. The combination of the intensely ionized streams representing the energy of Agni, and the powerful force driving the pumping action representing the power of Indra may be seen as the characteristic of Indragni. The junction point where the bar shape turns into the spiral shaped arms gives credence to the word Vishakha.

mitraḣ

Mi-tra is (so called) because he preserves (trayate) from destruction (pra-mI-ti) or because he runs (dravati) or measures things together (Mi), or the word is derived from the causal of (the verb) mid (to be fat). The following stanza is addressed to him. Yaskacharya quotes the following mantra to introduce Mitra Devatha.

mitro janānyātayati bruvāṇo. mitra dādhāra pṛthivīmuta dyām.
mitraḣ kṛšṭīranimišābhi cašṭe. mitrāya havyam dṛtavajjuhota

> Mitra leads with encouragement. Mitra supports the earth and the sky. Unceasing, mitra watches. Fat oblations are offered to Mitra. Mitra's role is that of a change catalyst and that of holding universal structures in their respective position. Mitra along with Varuna is one of the most prominent Adityas. Mitra and Varuna have complementary roles. The *Agastiya Samhita* text describes the construction of an electric battery whose cathode-anode pair is viewed as being controlled by Mitra and Varuna.

The nakshatra mantra for Anuradha is

ṛdhyāsma havyai namsopasadya mitram devam mitra deyanno astu.
anūrādhān, havišā vardhayantaḣ. ṣatam jIvema ṣaradaḣ savīrāḣ.
citram nakṣatramudagāt purastāt. anūrādhā sa iti yadvadanti.
tan mitra eit pathibhirdevayānaiḣ. hiraṇyayai vitatairantarikṣe

May we prosper, respectfully approaching god Mitra with oblations, with homage. Let a covenant of friendship (with him) be ours. Gladdening the Anuradhas (who are Mitra's Naksatra) with our oblation, may we live a hundred autumns with manly sons! The bright Naksatra which they call the Anuradhas, has risen in the east. Mitra goes to it by the ways that are the ways of the gods, the golden ways, extended in the atmosphere

An alternate meaning for the word *purastāt* is forward. The alternate translation of the phrase *hiraṇyayai vitatairantarikṣe* is an imperishable cluster in space. Mitra Devata is associated with the affinity principle. The word *dārā* refers to a flow away from the source and the word when reversed, indicates a flow back to the source[36] . Anuradha therefore refers to a quasi-gravitational effect. This interpretation also suits the nature of Mitra as the Devata of affinity.

---

[36]using the examples of *Charu* and *Ruc* and *Madhu* and *Dham* in the Nirukta text

There are two interesting facts about the region of the nakshatra Anuradha. This region includes the portion of the galactic central bulge which astronomers are studying intently. Gravitational forces here are strong due to the proximity to the galactic center and yet weak enough to allow stars and globular clusters to rotate independently around the galactic center. A delicate balance keeps collisions out of this compact space filled with rotating entities. The word Anuradha, defined as a force that pulls, may be pointing to this delicate balance among different gravitational fields in this region. Also of note is the presence of the "great attractor" which lies far beyond our galaxy in this direction of the sky. Mitra's affinity impulse is reflected in the gravitational pull of the great attractor on our galaxy. The mantras for Anuradha Nakshatra may be referring to two themes, namely, the well balanced paths taken by globular clusters around the galactic center, and the clusters of galaxies, moving gracefully towards a point beyond the great attractor.

indraḣ

Yaskacharya has presented many ways to derive the word Indra. Indra is so called because he divides food (irA + dr), or he gives food (irA + da), or he bestows food (irA+dhA), or he sends food (irA + dAraya), or he holds food (irA + dhAraya), or he runs for the sake of Soma (indu + dru), or he takes delight in Soma (indu + ram), or he sets beings on fire (indh). He is (so called) from doing everything says the Rishi *Agrayana.* He is so called from seeing everything (idam +dris),' says the Rishi *Aupamanyava.* Or the word is derived from (the verb) *ind*, meaning to be powerful, i.e. being so powerful he tears enemies asunder, or puts them to flight or he honors the sacrificer. The following mantra is addressed to him.

adardarutsamaŗjo vi khāni tvmarṇavānbabdadhānāgm aramṇāḣ.
Mahāntamindra parvatam vi yadvah sŗjo vi dhārā ava dānavam han.

> Thou didst pierce what issues out[37], create the channels, and atmospheric multitudes having water, and pressing each other hard[38] . O Indra, thou didst

[37] An original translation of ut-sa is spring. A better understanding in the context of astronomy is obtained by alternate derivations given in the Nirukta which are from ut-sr – moving upwards, or from ut-sad – raising upwards, or from ut-syand– flowing upwards or from the verb ud -to issue out.

[38]Original translation - rich in water, send them forth

uncover the mighty mountain, emit the outward flow, smite down the producer. The following other stanza is addressed to him.

yo jāta eva prathamo manasvāndevo devānkratunā paryabhūsat. yasya şmādrodasi abhyasetām nŗmņasya mahnā sa janāsa indrah

The wise god, who immediately on manifesting through the strength of his action protected all Devatas, at whose power (breath) heaven and earth tremble on account of the greatness of his might, he, O men, is Indra[39].

The Nakshatra mantra for Jyeshta is the following

indro jyešţhamanu nakşatrameti. yasmin vŗtran vŗtra tūrye tatāra.
tasminvaya-mamŗtam duhānāh. kşudhantarema duritim durišţhim.
purandarāya vŗšabhaya dhŗšņave.
ašādāya sahamānāya mīdhuše.
indrāya jyešţhā madhumadduhānā.
urum kŗņotu yajamānāya lokam.

Indra follows Jyeshta, the Naksatra. Under this Naksatra (this constellation), under which, in his fight with Vritra, he overcame Vritra, may we, milking the beverage of immortality, overcome hunger, misery, and failure in the sacrifice. Let Jyestha, who is milking the sweet oblation for Indra, the destroyer of strongholds, the fierce bull, the invincible one, the victorious one, the bountiful one,-give broad free space to the Sacrificer

Alternate interpretation of the above mantra shows Indra to be overcoming vritras in his own territory. The phrase *urum kŗņotu yajamānāya lokam* may refer to the expansion of the universe. This Mantra alludes to the conflict between negative and positive forces in which Indra's strength, vigor, and liberal nature secure him a victory after which he upholds the expansion of the worlds. *vŗtrā* is a representation of the negative forces which hoard precious resources. *vŗtrā's* influence is often seen in the clouds that hoard water from the world. The power of Indra vanquishes *vŗtrā* to bring down showers.

The word *jyešţhā* refers to something that is dominant. Indra's role in this Nakshatra fits well with this meaning. The creation of stars within a spiral galaxy winds down when its galactic nucleus becomes dull. An AGN (Active

[39] Nirukta proceeds to state that "thus is the thrill of a seer. His intuitive insight into reality makes him expresses the thrill with a narrative"

Galactic Nucleus) supplies plasma rich in ionized material to the molecular clouds in the entire galaxy. Scientists attribute the origin of the force, constantly driving the massive amounts of plasma, to a region in the immediate vicinity of a central black hole. Every spiral galaxy, such as ours, contains a black hole which has the power to gobble up all the star forming matter, depleting a galaxy of new stars. On the other hand, too much star formation, called rapid star bursts, occur with the galactic center pushing out too much star material. This can lock up the rich star raw material. A fine balance between the two scenarios is necessary.

Astrophysicists theorize that this balance is achieved by some force near the galactic core. This force is the strongest of all forces within the galaxy. The *jyeṣṭhā* region in the sky is close to the galactic black hole of our own galaxy. This region hosts the switch that controls star creation activity in the entire galaxy. The complex and powerful energies at play in this region represent Indra. Just as the seat of Indra in the human body is in the brain, his seat in the galaxy is next to the galactic black hole. Indra supports the forces of expansion that are ever victorious over the forces of dormancy.

nrṛtī

The word Mula means a root or a source. *nrṛtī* -is synonymous with destruction and calamity. The word is derived from (the root) R (to befall). Mantras to *nrṛtī* plead for a safe distance from her destructive influence. An example is Rig Veda Mandala 10, stanza 29. Here are a few lines from the mantra

pra tāryāyuḥ prataram navīya sthātāreva kṛtumatārathasya adha
cyavāna ut tavityartham parātaram sogm sāman nu rāye nidhiman
nvannam karāmahe su purudhaṣravāmsi tā no viṣvāni jaritā mamattu
parātaram sunirṛtirjihītām

> His life hath been renewed and carried forward as two men, car-borne, by the skilful driver. One falls, then seeks the goal with quickened vigor. Let Nirriti depart to distant places.

mo ṣu ṇaḥ Soma mṛtyave parā daḥ paṣyema nu
sūryamuccarantam. dyubhirhito jarimā sū no astu parātaram
sunirṛtirjhītam

> May we overcome our foes with acts of valor, as heaven is over earth, and hills are over lowlands! All these our deeds the singer hath considered.

The Nakshatra Mantra for Mula is –

mūlam prajām vīravatīm videya. parācyetu nrṛti parācā. gobhirnakṣatram paṣubhissamaktam. aharbhūyādyajamānāya mahyam. aharno adya suvite dadhātu. mūlam nakkṣatramiti yadvadanti. parācīm vācā nirṛtim nudām. ṣivam prajāyai ṣivamastu mahyam

May I obtain (the favor of) Mula (Root) (the Naksatra of Nirriti), (and) a progeny abounding in manly sons. Let Nirriti (the goddess of destruction) go away by a far path. May the Naksatra, united with cows and (other) domestic animals, be day (i.e. as bright as day light) for me, the Sacrificer. (Being) day (i.e. as bright as day light), let the Naksatra which they call Mula, establish us today in welfare. With my voice, I drive away Nirriti. Let prosperity be for my progeny, let prosperity be for me.

This Mantra introduces the idea that this area of the universe, itself a creation of *nṛtī*, is dear to her, and accordingly, her negative influence is limited here. There is an emphasis on the word *parāci*,"[40] meaning inward flowing. The mantra seeks her influence to flow inward to avoid annihilation (from her destructive character).

Black hole physics is an evolving science. It delineates an event horizon to be the boundary separating a region within which matter and light entering it from the surrounding space of normalcy disappear. Outside the event horizon a black hole exhibits the property of any heavy astronomical object. The general theory of relativity predicts space-time deformity near a black hole whereby the paths taken by a particle falling into a black hole appear to be bent towards it. At the event horizon of a black hole, the deformation is strong enough that no path leads away from the black hole. To an external observer an object nearing an event horizon appears to slow down taking an infinite time to reach it. *Nirriti* may be seen as the force beyond the event horizon and the description of Mula resonates with the properties of the space of stability just outside the event horizon. The anthropomorphic image of Nirriti in Purana texts is a fierce looking woman gobbling existence itself. The center of our galaxy is in the Mula, the Nakashtra region of the sky, meaning the root.

---

[40]The word *arvaacl* indicates in all directions.The word *praacl* can therefore refer to the exact opposite or as directed away inwards.

divyā āpaħ

Apah (waters) is derived from (the root) Ap (to obtain). The, following stanza is addressed to them.

Āpo hi šţhā mayobhuvastā na ūrhe dadhātana. mahe raṇāya cakṣase.

> Ye waters are indeed beneficent. As such bestow strength on us, so that we may look upon great happiness. Ye waters are indeed a source of comfort. As such bestow food on us, so that we may look upon great happiness, i.e. delight. āpa is what provides sustenance and comfort.

Apa in an emotional context is a reference to love. āpa in the spatial region is referred to as **divyā āpaħ.** Just as water is needed to grow food and to sustain the growth and nourishment of living entities, star material is needed for the growth of new stars in a galaxy.

The Nakshatra *ašāḍhāh* with *divyā āpah* as its Devatha has the following Mantra:

yā divyā āpah payasā sambabhuvuħ. yā antarikṣa uta pārthivīryāħ. yāsāmašāḍhā anuyanti kāmam. tāna āpah ṣagg syonā bhavantu. yāṣca kūpyā yāsca nādyāssamudriyāħ.
yāsca vaiṣantīruta prāsacīryāħ. yāsāmašāḍhā madhu bakṣayanti. tāna āpah ṣagg syonā bhavantu.

Let the heavenly waters, who have united with milk (i.e. with invigorating sap), those who are in the atmosphere, and those who are coming from the earth, let those Waters whose wish the Ashadas (i.e. the First Ashada, their Naksatra) obey, be pleasant, agreeable, to us. The Waters of the wells, and of the rivers, and of the sea, and of the ponds, and those Waters which are congealed, let those Waters, whose sweetness the Ashadhas (i.e. the First Ashadha, their Naksatra) enjoy, be pleasant, agreeable, to us.

An alternate interpretation for the mantra presents an astronomical idea clearly. To the fluids generated in the swelling [41] regions in space and with royal strength, Ashada is merely a reflection of their impulse. Let *āpah* carry softness and pleasantness, vibrating, gathering together, percolating, and sprinkling all

[41] *Payas*, commonly used to denote milk, is derived from the root pA -to drink, or from pyeya -to swell. Here the latter derivation provides a more meaningful translation

around. Let ashada obtain madhu, equated with Amruta/longevity, through them. Let these *āpah* become soft!

In the mantra for Jyeshta, the word ashada was an adjective of Indra. Here the same word becomes the name of the Nakshatra. The Nakshatra name also means invincible. This power has come from its Devata *divyā āpah* representing energy and matter that is swelling and flowing like a fluid.

The aShada and *JyeShta* Nakshatras hold positions on either side of the *Mula* nakshatra which hosts the black hole of our galactic center. Astronomically Ashada and Jyeshta share many common characteristics. However, the Nakshatra system highlights their differences. On the Jyeshta side, it highlights the force of Indra who keeps the galaxy humming with the birth of new stars. On the *aShada* side, it highlights the fluid/pliable aspects of flowing currents that reach into the galactic arms. Astronomically the central bar of the Milky Way galaxy is tilted away from the earth in the direction of Jyeshta and it is closer in the Ashada direction. There is thus a perspective difference between the two areas. The near end of the galactic bar feeds the Scutum-Centaurus arm. The Orion spur containing our solar system branches out of this arm. The Perseus arm branches out from the far end of the bar and connects to the other end of the Orion spur. The *aShada* region teems with clusters of red giant stars. The stars of this category mature quickly and explode into supernovae, constantly replenishing and enriching star material in this region. This star material then flows out into the galactic arms. The property of space in this region of the sky, which hosts material and energy that constantly reconfigure, mirrors the fluidity aspect of *divyā āpah.*

vişvedevāh

vişvedevāh is a grouping of Devathas. Nirukta does not analyze the word *vişvedevāh.* But according to mandala 10 of the Rig Veda, (Stanza 54) their overall characteristic appears to be an amalgamation of multiple Devatha energies. The combined stream of Devatha energies carried as *vişvedevāh* enable individual energies within the stream to fork off and subsequently manifest independently. Their skills make the generational existence of human beings possible.

mahimna eşām pitaraşcaneşire devā deveşvadadhurapikŗtum
samavivyacuruta yānyam şuraişam tanūsu vivişuh punah

> Part of their grandeur have the Fathers also gained: the Gods have seated mental power in them as Gods. They have embraced within themselves all energies, which, issuing forth, again into their bodies pass.
>
> sahobhirviṣam pari cakramū rajaħ pūrva
> dhāmānyamitāmimānaħ tanūšu viṣvā bhuvanā ni yemire
> prāsārayantapurudha prajā anu
>
> They strode through all the region with victorious might, establishing the old immeasurable laws. They compassed in their bodies all existing things, and streamed forth offspring in many successive forms.

The Nakshatra mantra for Ashada (Uttara) is

tanno viṣve upa ṣṛṇvantu devāħ. tadašāḍhā abhisamyantu yaňam.
tannakṣatram pṛthatām paṣubhyaħ.
kṛširvṛšṭir-yajamānāya kalpatām.
ṣubhrāħ kanyā yuvatayassapeṣasaħ. karmakṛta-ssukṛto vīryāvātīħ.
viṣvān devān havišā vardhāyantīħ.

So let the Vishvedeva listen to us. So let the Ashada (i.e. the Second Ashada, who are their Nakshatra) come to our sacrifice. So let the Nakshatra spread itself (i.e. spread its light) for the benefit of the domestic animals. Let agriculture and rain be favorable to the Sacrificer. Let the beautiful girls, the well-shaped young women, the powerful ones, who perform their work, and skillfully perform their work, let the Ashada (i.e. the Second Ashada), gladdening the Visve Devah with the oblation, come to the object of their desire, the sacrifice.

aShada Nakshatra here enhances the impulses of *viṣvedevāh* of regulating and of increasing showers and growth

Yagna is an imitation of the cosmic principle of evolution. The offering of clarified butter in a yagna imitates the principle of transforming matter (gross) into the energy (subtle). Additional substances being offered in fire represent frozen impulses within which are released into the energy domain. The release of energy impulses can trigger changes in the subtle planes of existence which eventually cascade as positive changes in the gross plane. The *aShada* region, as already discussed, is a galactic area containing clusters of massive stars.

These stars mature relatively quickly when compared to stars in other parts of the Milky Way. Their purpose appears to be to lead the cosmos towards higher concentration of heavier elements. The original cosmos contained only hydrogen and some helium. Over generations of stars, galaxies have produced other elements creating the mix of elements necessary for living entities to evolve. Our galaxy mirrors the transforming aspect of a Yagna, in a cosmic sense. *aShada* nakshatra highlights the evolutionary tendency of the cosmos. The influence of *viṣvedevāḥ* as the Devatas encapsulating other life supporting impulses can be seen at the junction point between the central region of our galaxy and the arms connecting to it. The near side junction point falls in the Ashada (Uttara) Nakshatra, which is ruled by the Visvedeva.

višṇuḣ

What is set free becomes Vishnu. Vishnu is derived from the root "vis" to pervade, or from "vy-as" to interpenetrate. Nirukta quotes a well-known Mantra

idam višṇurvi cakrame tredhā ni dadhe padam. samūdamasya pagmsure.

> Vishnu strode over this (universe). Thrice he planted his foot, enveloped in dust, and Vishnu strides over this and all that exists. Quoting Rishi *Şākapūni, yāskācārya* says - "thrice he plants his foot" stands for threefold existence. Quoting another expert Rishi *aurṇavābhā* states that the three footprints are on the mountain of sunrise, on the meridian, and on the mountain of sunset. The foot-print is not visible in the stormy atmosphere. It appears from a reference to "dust" in the Mantra or it may be used in a metaphorical sense, i.e. his footstep is not visible, as if enveloped in a dusty place. The word *pagmsuvaḣ* means something that lies scattered below, or is trodden down.

It is clear that even at the time of the writing of the Nirukta, different schools of thought viewed the phrases such as "three foot prints" differently. The Purana texts take an anthropomorphic approach in highlighting the wide stride impulse of Vishnu by seating him on an Eagle.

Şṛoṇa (Shravana) is derived from şṛ -to hear. The Devatha of this Nakshatra is Vishnu. The associated Mantra is

şṛuṇvanti şṛoṇā-mamṛtasya gopām.
amṛtaamṛta. punyāmasyām upaşṛṇomi vācām.
mahīm devīm vişṇupatnīmajūryām.
pratīcīm enāgm havişā yajāmaḣ. tredhā višṇu-rurugāyo vicakrane.
mahīm divam pṛthvī-mantarikşam. taccroṇaitişrava iccamānā.
puṇyagg şlokam yajamānāya kṛṇvatī.

They (the people, or the holy men) hear Ṣṛoṇa, the guardian of the beverage of immortality. I hear her holy voice. To the great goddess (Sroni), Vishnu's wife, who is not subject to old age, we offer the oblation, as she is turned towards us. Triply wide-paced Vishnu strode through the great sky, the earth, and the atmosphere. So Sroni (who is his wife and his Naksatra) goes, wishing for fame, creating pure glory for the Sacrificer.

There is a notable use of the word *Vaac* in this mantra which the Nirukta has analyzed. An alternate interpretation of the word *pratici* was already discussed with the Nakshatra Mula. It conveys the idea of an inward directed flow which narrows down the astronomical phenomenon in this region of the sky to specific objects. The model of trifold separation of the cosmos, discussed in the earlier chapter, is highlighted by the phrase *mahīm divam pṛthvī-mantarikṣam.* The phrase *taccroṇaitiṣrava iccamānā* can be interpreted as those desiring material needed for growth[42] .

We look for an astronomical phenomenon that highlights the model of the trifold separation of the cosmos. Cygnus X-1 is a good representation of this phenomenon where a very dense black hole at the center is the "Bhu" element, surrounded by an "Antariksha" element in the accretion disc, and the "Dyu" element in the radiant energy, flooding far out to the edge of the universe. The "Bhu" element in this representation of the model also incorporates the "inward flowing" property which the mantra refers to. The word "Vaac" used in this Mantra may be drawing our attention to the auditory perception. This seems appropriate because there is no visual perception associated with a black hole. Cygnus X-1 is a source of some of the highest energy radiations seen in the Milky Way. The far reach of this radiation fits well with the phrase "wide stride" associated with the Devata Vishnu. The inner Space of a black hole transcends time, as according to science, timespace ends at its event boundary. Nakshatra Shrona is the keeper of immortality.

vasūḥ

yad vivāsate sarvam - Vasus are so called because they put on everything. Agni gets the name Vasava -- being associated with the vasus. Indra is also

[42] *Shrava* in Niruktha is translated to mean *Anna* which is often translated as food. Here it is considered to be figurative – any material that is needed for growth

called Vasava because of his association with the vasus. On account of shining forth, the rays of the sun are also called Vasus. Elaborating on this idea, Nirukta proceeds with the following Mantra.

sugā vo devāh sadanamkarma ya ājagmuh savanamidam jušānāh jakşivāgmsah papivāgmsaşca vişve'so dhatta vasavo vasūni -

The dwelling has assembled well with your arrival. With joy participate in the *Savana* ritual (see description of the 360 day Savana ritual in Chapter 3, Nakshatra Shravishta is also associated with the controversial way that Vedic astronomy has been dated). Having tasted let Vasus bestow treasures. The following stanza, is addressed to them.

jyamā annam vasavo ranta deva urāvantarikşe marjayanta
şubhrāh arvākpatha urujrayah kŗņudhvam şrotā dūtasya
jagmušo no asya.

The divine Vasus finding joy in the food from jyam-earth, shining in space, in a large body, clearing impurities, let them follow into the deep space within[43] to increase vigor. Listen to this our messenger[44], who has started on his journey. A specific set of Devatas manifesting within a specific form for a certain duration may be called the vasus because the name Vasus is derived from the root "vas" - to dwell. Vasus are an embodied aspect of selected Devatas. For example, the form of a human being is considered to be held together by eight Vasus.

The Nakshatra mantra for Shravishta is

aşţau devā vasavssomyāsah. catasro devī-rajarāh şravişţhāh.
te yaňam pāntu rajasah parastāt. samvatsarīņa-mmŗtagg svasti.
yaňam nah pāntu vasavah purastāt. dakşiņato'bhiyantu şravişţhāh.
puņyannakşatramabhi samvišāma. mā no arāti-raghaşagmsā'gann.

Let the eight divine, Soma-loving, Vasus, and the four divine, undecaying, Shravisthas (who are their Naksatra) protect, far away from dust, the sacrifice, (that is) yearly, imperishable prosperity. Let the Vasus protect our sacrifice in the east; let the Shravisthas (who are their Naksatra) come from the south. Let us meet the pure Naksatra. May wicked hostility not reach us!

---

[43]*Pravak* which is defined as moving outwards. Therefore *arvaak* here has been translated into deep within

[44]In the Nirukta – Agni is called as the messenger who has started on his journey

The connection of Vasus with Soma, the phrase "changeless şhravişţhā," the adjective longevity and the epithet "dust free" seen in this Mantra point to something existing for a long time. The word *şravişţhā* being from the word root şŗ "to glide/flow" along with the suffix şţhā implies something that is in a steady flow.

We look for objects that have been long time denizens of the universe to qualify for the adjective "longevity" and also far away from the bustle in the active areas of the galaxy to fit the phrase "dust free". The object is also something in a constant motion as per the interpretation of the word root of the name of this Nakshatra. All these match the Globular clusters which are the oldest occupants of our Milky Way galaxy. They move far out in the halo region of our galaxy, only loosely tied to the movement of the stars close to the disc of our galaxy. A globular cluster contains a vast variety of other astronomical objects, including blue stranglers, pulsars, x-ray binaries, planetary nebula, and neutron stars. Some globular clusters may even include a black hole at their centers. Vasu Devatas are a grouping of eight well known Devatas who are active within a defined space sustaining a specific form that envelopes that space. Globular clusters, each hosting close to a hundred thousand stars in this Nakshatra region may be identified as signature entities of Vasu Devatas and the associated Nakshatra. Astronomers say that the four globular clusters -- M15, M2, M30, and Segue -- will likely rotate around the Milky Way center for millennia to come.

varuņaḣ

Varuna is so called because he covers, **vŗ**. The following stanza is addressed to him.

nicinabāram[45] varuņaḣ kabandham[46] .pra sasarja rodasi antarikşam
tena vişvasya bhuvanasya rājā yavam na vŗşţirvyunatti bhūma.

> Varuna sent forth the clouds, opening downwards, and created heaven, earth, and the intermediate space. With it, the king of the entire universe moistens earth as rain, the barley. The following, another stanza, is addressed to him.

---

45 The alternate translation is door opening downwards

46 *kavanam*, which means water, is deposited into it. The verb bandh is used to denote an unfixed state. It is comfortable and unrestrained

tamū pu samanā girā pitṛṛṇām ca manmabhiḣ nābhākasya praṣastibhiryaḣ sindhūnāmupodaye saptavasā sa madhyamo nabhantāmanyake same.

I praise him equal to the manes with the panegyrics of nābhāka[47] , who possesses seven sisters at the birth of those that flow , and who belongs to the middle region, Let him force others to burst open to the same degree. Varuna as a controller of waters is a cosmic impulse causing fluids/matter to spread out.

The Mantra to *ṣatabhiśaj* begins with praise to Varuna as having supremacy over *kṣatra* (regions) and to *ṣatabhiśaj* as having supremacy over *na-kshatras.*

kṣatrasya rājā varuṇo'dhirājaḣ nakṣatrāṇagm
ṣatabhišagvasiṣṭhaḣ tau devebhya kṛṇūtau dīrghamāyuḣ.
ṣatagm sahasrā bhešajāni dhattaḣ. yaṅanno rājā varuna upayātu.
tanno viṣve abhi samyantu devāḣ. dīrghamāyuḣ pratiradbhešajāni.

The king of the Kshatriyas, the sovereign king Varuna, and Shatabhishaj (who is his Naksatra), the best of the Nakashtras, these two create long life for the gods; (for) they give (those) one hundred thousand medicines. Let the king Varuna come to our sacrifice; let all the gods come together to our sacrifice. May the Naksatra Shatabhishaj, pleased with it, give us long life and medicine!

The Nakshtra mantra emphasizes Varuna's influence in this Nakshatra as bringing about recovery with his ability to spread out materials. The very name of the Nakshtra, *ṣatabhiśaj* implies a healing nature that manifests multifold.

Water has a tendency to spread out and cover an area. Water is found in abundance in the oceans which covers the whole planet. Vedic texts attribute to Varuna the impulse of spreading and occupying. Since water spreads, Varuna is referred to as the ruler of waters. Hydrogen is the most abundant and widely spread matter in the cosmos. Interstellar hydrogen is Varuna's signature in space. The affinity impulse of the Devata Mitra complements the spreading impulse of Varuna, and together, they keep the universe in equilibrium. We look for a region of space dominated by hydrogen, the carrier of Varuna's signature in space. The spherical halo of our galaxy extending 300,000 light years outwards and filled with extremely low density hot hydrogen gas may be one representation of Varuna's impulse. As this halo is not restricted to only the direction of Nakshatra Shatabhishaj, we look for

---

[47]According to Nabhaka, Varuna having seven sisters at birth, of those that sprinkle (*syandanam*) is called to be of the middle region

another astronomical object that highlights the tendency to spread out. NGC 7482, a flattened globular cluster, resides some 100,000 light years away from Earth. Scientists attribute its flat distribution to the fact that most of the original stars from its outer radius of lighter mass have dispersed, leaving behind the remaining stars of similar mass. NGC 7482, in the direction of the Nakshatra Shatabhishaj region, can draw our attention to the equalizing nature of Varuna.

Subtleties of Space

aja ekapād

"ekam pādam notkhedati" -- he who does not draw one foot out is the Vedic quotation for Aja Ekapad. He is Surya. The word aja means "unborn". This word is also a common reference for a goat. The word ekapad means the one-footed driver, or he protects with one foot, or he drinks with one foot, or he has only one foot. The word Surya, who some consider the same as aja ekapad, is derived from the word root sṛ -to move, or from su -to stimulate, or from svir -to promote well. The following stanza is addressed to Surya.

udu tyam jātavedasam vahanti ketavaħ. dṛṣe viṣvāya sūryam.

> Rays uplift him, the god who has all created things as his property, i.e. Surya, for all to see. The following, another stanza, is addressed to him.

citram devānāmudagādanikam cakṣurmitrasya varuṇasyāgneħ
āprā dyāvāpṛthvī antarikṣam sūrya ātmā jagatastasthušuṣca.

> The variegated splendour of the gods, the eye of Mitra, Varuna, and Agni, has gone up. He has filled heaven, earth, and the intermediate space. Surya is the soul of the moving and the stationary.

The Nakshatra Mantra for Proshtapada (Purva) is

aja ekapādudagātpurastāt. viṣvā bhūtāni pratimodmānaħ.
tasya devāħ prasavam yanti sarve. prošṭhapadāso amṛtasya gopaħ.
vibhrājamāssamidhā na ugraħ. Ā'ntarikṣamaruhadagandyām.
tagm sūryam deva-majamekapādam. prošṭhapadāso anuyanti sarve.

Aja Ekapad (the one-footed he-goat, i.e. the Sun) has risen in the east[48], greeting with joy all beings. All the gods, (and) the Proshtapadas (i.e. the First Prosthapadas, who are his Naksatra), the guardians of the beverage of immortality, follow his impulse. Shining, flaming, powerful, he has mounted the atmosphere, and has reached the sky. All the Prosthapadas (i.e. the First Prosthapadas, who are his Naksatra) follow him, the Sun god, Aja Ekapdd.

The phrases shining fiercely, risen to the top, all Devathas following lead us to an understanding of the nature of this Nakshatra. The word Proshtapada contains the affix "pada." This affix is also seen in the name of the associated Devata. The word Surya, used in the Mantra is as an epithet for Ekapad.

One derivation of the word Proshtapada is from the word *pṛṣ̌ṭha* -- something standing forth prominently or elevated. It guides us to look for a semblance in this Nakshatra region. The tilt between the galactic and stellar planes causes the galactic plane to reach a northern most point in the Proshtapada region of the sky. A contradiction lies in the meaning of the name of the Devata Aja Ekapad. The Vedic Rishis draw attention to an aspect of the cosmos through this contradiction related to the mobility of a biped. A walker cannot advance forward without placing a second leg forward, but the word Ekapad suggests such a possibility. The only entity in the cosmos which can be thought of as representing this contradiction is space. Space remained even while the primordial plasma moved out of it and created the visible universe. We look for a peculiarity of space to explain the Aja Ekapad impulse. Astronomers have made a puzzling observation related to the direction of the spin of galaxies: the preponderance of galaxies with counter-clock wise rotation can only be explained if the universe retained some memory from the past; only space itself could have retained it! We then search for a specific astronomical object that also exhibits a peculiarity related to its spin. An instance of this is the central bulge of the galaxy NGC 7331, often termed the twin of the Milky Way galaxy. In spiral galaxies, the central bulge co-rotates with its disk but the bulge in the galaxy NGC 7331 rotates in the direction opposite to the rest of the disk. Nirukta considers the Sun to be the Aja Ekapad. This is true as all movements on Earth are attributed to the Sun, static at the center of the solar system.

---

[48]The alternate meaning of *Purastat* is "from earlier" and appears to provide a better clue to identifying an astronomical object

ahir budniyaȟ

A cloud is called an ahi on account of its motion. Another meaning of ahi is serpent. Ahi the serpent can be derived from the same root as Ahi indicating a motion. However Ahan – to attack is a better derivation for ahi, the serpent. Ahi in this case appears to be related more to a word derivation implying motion. Ahi moves in the atmosphere. Budhnam is where waters are held or bound. It is therefore a reference to the atmosphere. Another meaning of the word budhnam is the body. It is budhnam because the breath is held, bound in it. Budhnam may also refer to space where energy is held. Elaborating on this, Nirukta explains:

Abhāmukthairahi gŗṇīše budhne nadīnām rajassu pīdam

> In your abode amidst the excitement in the flowing currents you take in whatever is released by a water derived source. The following stanza is addressed to him.

mā no'hirbudhnyo ripe dhānmā yaňe asya stridhdŗtāyoȟ

> May Ahi who dwells in the atmosphere not cause damage! May the yagna not fail!

The Nakshatra mantra for Proshtapada (Uttara) is

ahirbudniyaȟ prathamā na eti. şrešţho devānāmuta mānušānām. tam brahmanāsSomapāssomyāsaȟ.
prošţhapadāso abhirakşanti sarve. catvāra ekamabhi karma devāh.
prošţhapadā sa iti yān vadanti. te budhniyagg stuvantaha.
ahigm rakşanti namasopasadya.

> Ahi Budhniya (the Serpent of the depth) wanders, spreading himself, he, the best of the gods and of men. The Brahmans, who drink the Soma and who love the Soma, and all the Proshtapadas (i.e. the Second Proshtapadas, who are Ahi Budhniya's Naksatra) protect him. The four gods, whom the people call the Proshtapadas (i.e. the Second Proshtapadas), go to one and the same work[49]. Praising Ahi Budhniya, who is worthy of worship, they protect him, reverently approaching him with homage

[49] *Karma Devatas*, four in number, need to be understood in the Vedic context instead of the word to word translation given here.

The four Karma Devatas protect Ahir Bhudhnya. An aspect of nature both dominant and fragile is indicated by this Mantra.

The Vedic Pundits use a popular mantra[50] to seek the help of Ahir Budhnya to protect the mantras in their memory or consciousness. The Devata who can provide such protection himself requires to be protected! Mantras to Ahir Budhnya mostly include a reference to Aja Ekapad. These facts indicate Ahir Budhnya to be a deeper aspect of consciousness, which, in the Vedic tradition, is a property of space. Does the Aja Ekapad impulse give creation its gross mobility? Does the Ahir Budhnya impulse cause the subtle movements and vibrations? The phrase "delicate movement from a depth within a still environment" provides a clue to a physical representation of the Devata Ahir Budhnya in the sky. The closest phenomenon may be a Mercury-Manganese star. Alpha Andromedae is the brightest of Merucry-Manganese stars and lies in the Proshtapada region. As the name suggests, this star's spectrum is rich in mercury-manganese and other elements. Its rotation is relatively slow, and as a consequence, its atmosphere is relatively calm. Within it, atoms of some elements sink under the force of gravity, while others are lifted towards the exterior of the star by radiation pressure.

Miscellaneous Devatas

pūšā

Nirukta says that when the sun sees an increase of rays, he is called *Pusan*. The following stanza is addressed to him.

şukram te anyadyajatam te anyadvišurūpe ahanī dyaurivāsi. vişvā hi māyā avasi svadhāvo[51] bhadrā te pūšanniha rāti[52]sastu.

> Thy one form is bright (lohitam), thy other is yajatam (adorable – moon) causing night and days to be irregular[53].You protect the insights[54] into the universe. Here let thy gifts be blessed, Pusan, rich in food. The following, another stanza, is addressed to him.

---

[50] *Ahir Budnya mantram me gopaya*

[51] Explained as inherent power/own position

[52] Alternate meaning is generosity

[53] Nirukta explains višurupe to be višamarupe te ahani karma dyau rivacāsi

[54] Nirukta explains māyā here to be sarvāṇi prañanānyavasi

pathaspathaḣ paripatim vacasā kāmena kṛto abhyānaḷarkam.
sa no rāsaccurudhaşcanrāgrā dhiyamdhiyam sipadhāti pra
pūšā

Let the sun, the master of all paths, made ready to be known through desire and speech, bring the foremost[55] gifts born out of precession[56]. Let him make every action of ours to be fruitful.

The following is the Nakshatra Mantra for Revati

pūšā revatyanveti panthām. puşţhi patī pašupa vājavatyau.
imāni havyā prayatā jušāṇa. kşudrān pašūn rakşatu revatī naḣ. gāvo
no aşvāgm anvetu pūšā. annagm rakşantau bahudhā virūpam. vājagm
sanutām yajamānāya lokam.

Pusan and Revati (who is his Nakshatra) follow the path (leading to our sacrifice) - (they are) the two sovereigns of prosperity, the protectors of domestic animals, the possessors of dwellings full of wealth. Pleased with these offered oblations, let them come to our sacrifice by easy ways. Let Revati protect our small domestic animals; let Pusan follow (i.e. look after) our cows and horses. Protecting food, which is various in many ways, let them both grant to the Sacrificer wealth, and the sacrifice.

The Mantra highlights the connection between Pusha and the paths. The rest of the mantra also emphasizes the connection with nourishment -- a liberal supply of and the protection of. The word Revati as derived from the root *rI* is to release/bestow. It refers to something that bestows and is consistent with the epithets to Pusha as a liberal supplier of nourishment.

The Vedic concept of time, namely, the Manu-Kalpa model, conveys knowledge of the precession of the equinoxes. The zero point of the sidereal coordinate system in Jyotish falls within this Nakshatra. It is logical to look for references to the phenomenon of precession. A mantra of Pusha quoted in the Nirukta contains the word Ayana, which in astronomy texts refers to precession. The connection between precession and Pusha/Revati is suggested by the phrases in the Nakshatra mantra - varying day/night, path after path - *pathaspathah* and the controller of circular movement *paripathim*.

---

[55] Foremost – Nirukta provides additional clarity on what is here translated as formost - invigorating draughts of sparkling surface, treasures of noteworthy surface

[56] *sa no dadhātu cāyanīyāgraṇi (ayanīya agrani* – first gifts of an ayana) *dhanāni*. This clarification in Nirukta refers to Ayana - precession

Pusha is a name for the Sun with increased rays. Increased rays may refer to longer day times. The longest periods of sunlight are observed in the polar region. The Polar region also has extreme swings in the lengths of days/nights. Longer summer days within the polar circle lead to growth and bountiful vegetation. A star falling within the Revati Nakshatra and guiding the observation of precession can be an astronomical signature of Pusha. Alpha Cephei, relatively close to the earth, at a distance of only 49 light years is located almost exactly along the path traced by the Earth's North Pole. It periodically becomes a pole star with an offset of 3 degrees. This will recur around the year 7500 CE. It currently serves as Mars' pole star.

aşvinau

Ashwins are the first of the many celestial types of Devatas analyzed in the Nirtukta. The Ashwins are so called because they pervade (as) everything, one with moisture (rasa), the other with light (jyothi). According to *Aurnavabha* they are called Ashwins on account of their horse ownership (asva). Many opinions have been captured in the Nirukta about the twin nature of Ashwins - heaven and earth; day and night[57]. The following stanza is addressed to them.

vasātisu sma caratho'sitau petvāviva kadedamaşvinā yuvāmabhi devāgm agaccatam

> You wandered like two black clouds during the nights. O Ashwins, when was it that you came to the gods? Nirukta states that the meaning of this mantra is obvious! From this statement it appears that the phenomenon represented by Ashwins may be a frequent or common occurrence. Ashwins are mostly praised together. In a rare mantra they are individually addressed as Nāsatya, the other as the son of ušā. The following, another stanza, is addressed to them.
>
> iheha jātā smavāvaşītāmarepasā tanvā nāmabhiȟ svaiȟ -
>
> Born here and there they two are praised together with reference to a body unstained by sin and to their own names. "jišņurvāmanyaȟ sumukhasya" – one is of incredible strength "sūrirdivo anyaȟ subhagaȟ putra ūhe" the other is a Sun .

[57] The aspect in the dark - *anutama* is the middle regions, and the aspect in the light is the sun.

pratаryujā vi bodhayāşvināveha gaccatām asya Somasya pitaye

Awaken the Ashwins, who yoke early in the morning." May they two come here to drink of this Soma!

The highlights of the Nakshtra mantra for Ashwayuja are – sensitivity, easy to direct, healers, messengers, filled with pleasantness.

tadaşvināvaşvayujopayātām. şubhamgamiştḥau suyamebhiraşvaiḥ.
svan nakşatragm havişā yajantau
madhvāsampŗktau yajušā samaktau.
yau devānām bhišajau havyavāhau.
vişvasya dūtā-vamŗtasya gopau. tau nakşatram jujušāņopayātām.
namo'şvigbhāyām kŗņumo'şvayugbhyām

So, along with Ashwayuja (i.e. with the two Ashwayuja, who are their Naksatra), let the two Ashwins, swiftly driving on their shining path, come (to us) with their tractable[58] horses,-(let them come) offering the oblation to their Nakshatra, after having been satiated with honey, and adorned with the sacrificial formula. Those two who are the physicians of the gods, the bearers of the oblation, the messengers of the universe, and the protectors of the beverage of immortality,-let them both, delighted in their Naksatra, come (to us). We pay homage to the two Ashwins, (and) to the Ashwayujas (their Naksatra).

The word Ashwayuj refers to swift movement. The meaning of the word Yuj is yoked/linked. The word ashwa can indicate anything swift, such as a horse. The Beta Arietis binary star system has a rapidly spinning primary star and a fainter secondary star. They are located in the Ashwayuja Nakshtra. Scientists detect the Doppler shift from this pair and, using shift data, have obtained important clues about the mass and type of stars. Astronomers have mapped most stars in our galaxy through this procedure. Vedic mantras praise the Ashwini pair of Devatas as the messengers of the universe.

The principle of conservation of the angular momentum favors formation of multi star systems such as a binary pair, an idea discussed in detail in the following chapter. The Nakshatra mantra for Ashwayuja, which includes phrases such as "easy to guide," may convey the astronomical principle of conserved angular momentum during the process of star formation.

[58]The interpretation of the word suyamebhi is easy to guide

yamaḣ

Yama is from the root yam- to restrain. The following stanza is addressed to Yama

pareyivāsam[59] pravato[60] mahīranu bahubhaḣ panthāmanupaspaşānam[61]

releasing gently from all paths of the world– .

vaivasvatam samgamanam janānām yaman rājānam havišā duvasya[62].

Son of the Sun, the final confluence point for men, the royal Yama we seek with the offerings of havish. Agni is called Yama also. Yama is explained to be Agni by the following mantra

saneva sŗţhāmam dadhatyasturna diduttvešapratikā

Like a manifested missile either terrorizing or instilling courage (based on the view point) as does an arrow's appearance indicating it to be dreadful or bright. Yama and Indra are considered as twins with their mother being here and there – "yamo ha jāta indreņa saha samgataḣ yamāviheha mātarā."

The word *apa bharaņī* is formed by the words *apa* (off/away/back) and *bharani* derived from bearing/maintaining. This Nakshtra name conveys the role of its Devata, Yama, of releasing one from the ups and downs of life in the material plane. This release may be considered either as a gift or as a punishment based on one's view point. The Nakshatra Mantra

apa pāpmānam bharaņir bharantu.
tad yamo rājā bhagavān vicaşţām. lokasaya rājā mahato mahān hi.
sugam naḣ pantham abhayam kŗņotu.

---

[59]Using the explanation in Nirukta *paryāgatavantam* – revolving state

[60]Using the explanation in Nirukta *udvato nivata iti* – ups and downs

[61]Using the explanation in Nirukta *pāşayamānam* – binding

[62]Using the explanation in Nirukta *duvasyati- āpnoti karma* - gaining

yasminnakṣatre yama eti rājā. yasminnena-mabhyaśimcanta devāḥ.
tadasya citragm[63] haviśā yajema. apa pāpmānam bharaṇīrbharantu

Let the Bharanis (who are the Nakshatra of Yama) take away our evil; let the blessed king Yama look after that, for he is the great king of the great world. Let him make our way easy and safe. To that Nakshatra under which king Yama wanders, to that Nakshatra under which the gods did consecrate him, to that bright Nakshatra of him, let us offer the oblation. Let the Bharanis take away our evil.

This Nakshatra highlights the act of carrying away the bad and providing protection on a path that is easy. Yama dominants this region. Other Devatas have appointed him to be the ruler of this Nakshatra

Yama, like the Pitru Devatas, refers to the phenomenon of existence and termination. Yama carries away a man's life force from the observable world, revolving through its varied vicissitudes, to the unseen world. An astronomical idea that may fit the description of the world of Yama could be a region that exists contrary to the vibrancy and energetic aspects of the universe filled with life supporting impulses. Regions of vibrant energy go through constant changes -- matter forming new types of atoms and molecules transforming from one kind to another. Yama's region would not have these changes. A star such as the Sun comes under the influence of Yama when fusion ceases and shuts down life forms in the stellar system.

In stories from the Puranas, Yama is the rebel son of Vivasvan, another name for the Sun. Side by side regions of dormancy of energy, representing Yama, and regions of dominance of energy, representing Indra, corroborate the concept "Yama is a twin of Indra and whose mother is here and there." Voids and super voids are astronomical regions that lie in utter contrast to the observable regions of the universe. Scientists postulate these regions to be filled with dark matter and dark energy. These regions lie beyond the observable realms of the universe. The Eridanus super void is among the largest of such areas, and it falls in this Nakshtra region. It is a good

---

[63] *Chitra* in this mantra appears to correspond to the *Chitragupta* impulse of Yama. Chitragupta indicates something with a distinct signature but yet beyond the reach of traditional methods of observations. Rig Veda Mandala 8, Hymn 21, Stanza 18

representation of the ideas which the Nakshatra Apa Bharani and its Devata Yama convey.

The Full and the New Moons

Two additional Mantras are included as a part of the set of Nakshtra Mantras. Their relative position throws light on the logic behind the sequence. Apa Bharani is followed with the Mantra for *Amavasya* or the new moon, representing darkness. The mantra for the full moon is located half way through, after Vishaka Nakshatra whose Devatha is Indra-agni, a very intense form of energy. The 30th Mantra is dedicated to Brahma. Abhijith Nakshatra, whose Devata is Brahma is the 28th Nakshatra. Vedic astronomers did not assign a measurable region of the sky to this Nakshatra; it exists at a junction between Shravana and Ashada.

Devatas are classified as aspects of universal energy in order to explain the observable phenomenon of creation. In the Vedas, the creation does not exist separately from its creator – Brahma.

8. *Bridging the Gap*

The space between stars in the night sky appears empty to the casual observer, but large expanses are permeated with gas, primarily hydrogen, and traces of oxygen, nitrogen, and carbon spread thinly and unevenly. Denser regions containing larger molecules can be a thousand times more packed than the average distribution, with distant stars appearing dim due to the accumulated grains of dust blocking starlight. Astrochemists presently identify in excess of one hundred different molecules in the interstellar medium, their respective presences often revealed by spectroscopic signatures in radio frequencies. Large radio telescopes were not available prior to the 1970s, but today their widespread utility aids astronomers in the study of the structure of the galaxy and in the formation of plausible theories on star formation.

Interstellar gas clouds can contain enough raw matter to create millions of new stars. The creation process begins when a heavy clump of interstellar gas collapses under its own weight. The maintenance of a relatively low temperature is imperative during the early phases of collapse; the steady rise of pressure in the medium can abort the process.

Interstellar gas is generally just mere degrees above absolute zero. Molecules colliding under the cold conditions rotate to emit radiation with energetic collisions occurring more frequently when two clumps meet. Any radiation is fundamentally a loss of energy from the system and helps maintain low cloud temperatures. Increased radiation generally indicates progression towards star formation, and astronomers can identify regions of intense activity through corresponding radiation.

Having evolved from molecular clouds[64], stars are bound by the forces of gravity. A gravity-bound association of stars is called a cluster, with astronomers categorizing clusters as "open" or "globular" according to weak or strong gravitational attraction. Over one thousand open clusters are known to exist in the Milky Way galaxy. Due to weak gravitational binding, approximately half the stars of a new open cluster can be expected to drift away in a few hundred million years (de La Fuente, 1998). The Pleiades, also called

---

[64] An average size molecular cloud has enough raw material to form a few thousand stars

M45 in the Messier Index (Stoyan), is a famous open cluster, prominently visible in the night sky due to its proximity and domination by hot blue stars. The Pleiades spans nearly 2 full angular degrees in the night sky, almost four times the size of the full moon. Most of its stars formed in the last hundred million years.

But hot blue stars, noted for high surface temperatures and luminosities, comprise just a tiny fraction, 0.00003%, of stars in the Milky Way. Astronomers classify stars known on the main sequence[65] into the table below (Harvard Spectral Classification).

	Surface Temp	**Star Appearance**	**Radius**	**Luminosity**
O	≥ 33,000	Blue	≥ 16 $M_\odot$	≥ 30,000 $L_\odot$
B	10,000–33,000 K	blue white	2.1–16 $M_\odot$	25–30,000 $L_\odot$
A	7,500–10,000 K	white to blue white	1.4–2.1 $M_\odot$	5–25 $L_\odot$
F	6,000–7,500 K	White	1.04–1.4 $M_\odot$	1.5–5 $L_\odot$
G	5,200–6,000 K	yellowish white	0.8–1.04 $M_\odot$	0.6–1.5 $L_\odot$
K	3,700–5,200 K	yellow orange	0.45–0.8 $M_\odot$	0.08–0.6 $L_\odot$
M	≤ 3,700	orange red	≤ 0.45 $M_\odot$	≤ 0.08 $L_\odot$

[65] Stars generating light by the fusion of hydrogen are considered to be in the main-sequence

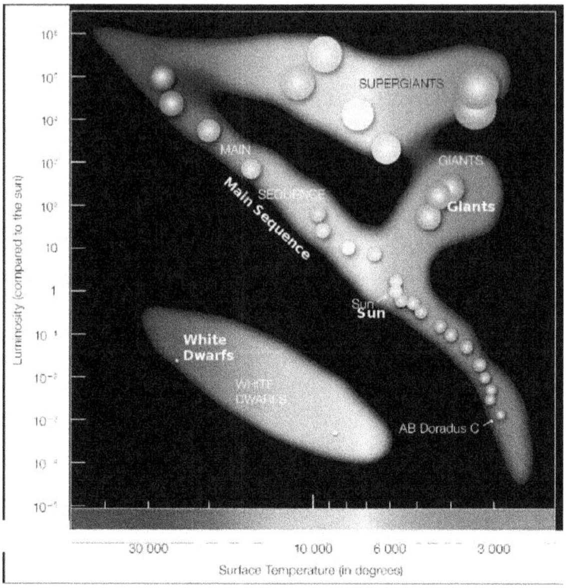

Figure 20 : Hertzsprung-russell diagram of star classification

Pleiades

Astronomical Phenomenon – Open Cluster
Nakshatra Characteristic – piercing, intense, fire, energy

Astronomers believe our own Sun formed in an open cluster like Pleiades and migrated to its present position. A member of an open cluster becomes dislocated due to tidal forces associated with the gravitational field of a passing system of stars. Infrared pictures (Figure 21) taken by NASA WISE show the cluster is surrounded by a huge cloud of dust transited by its stars. *Agni,* the general energy principle of the Vedas, is present throughout the universe but intensely so in this region. The intensity of energy in the cluster, due to the presence of hot blue stars and clouds of dust acting as smoke covering a flame, became equated in the cognitive awareness of the Rishis to the principle of *Agni.*

Figure 21: Pleiades region of the sky credited to NASA/JPL-Caltech/UCLA

Figure 22 : Orion Nebula, Picture Credit – NASA/ESA

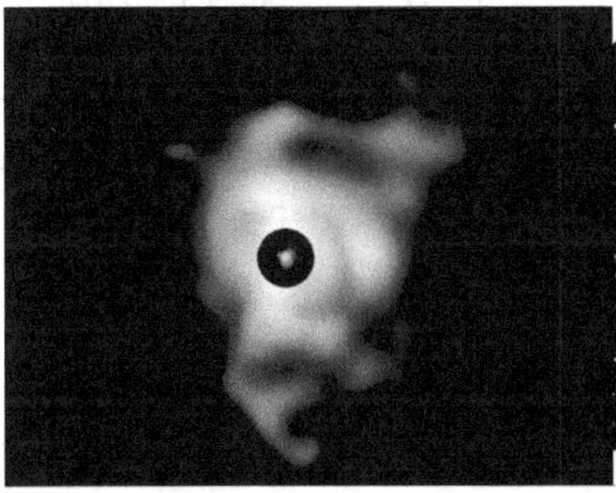

Figure 23 : Betelgeuse as the dark circle at the center with the complex atmosphere surrounding it. Picture credit ESO VLT

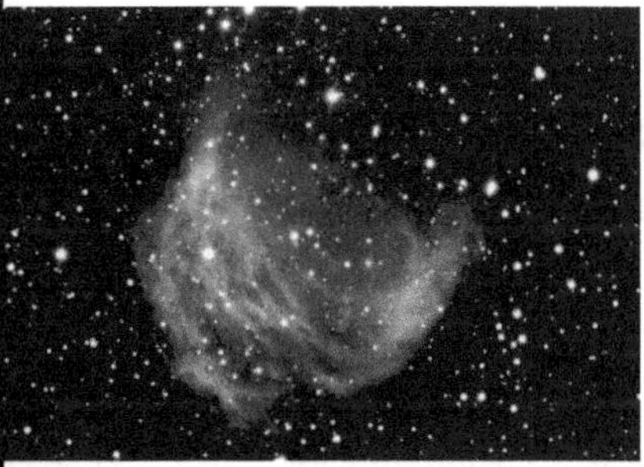

Figure 24 : Abel 21, Picture Credit: 24 inch telescope on Mt. Lemmon, AZ. Courtesy of Joseph D. Schulman

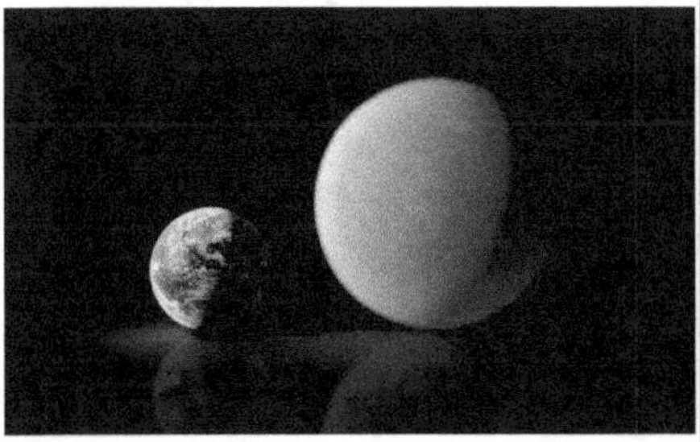

Figure 25 : Terrestrial planet in Cancri 55 system compared in size to the Earth

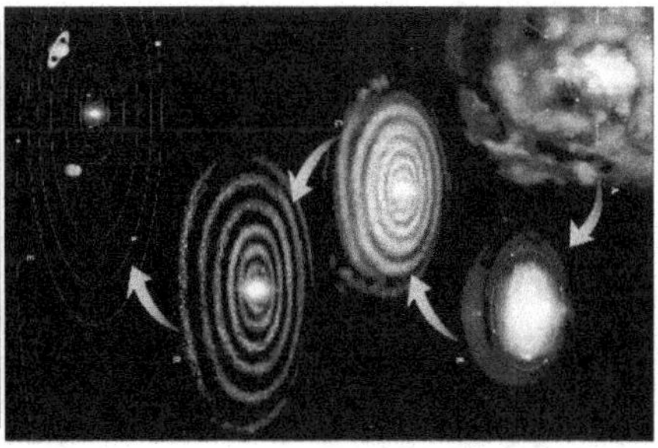

Figure 26 : Steps in the formation of a planet Picture credit: Plymouth State University

Aldebaran

Astronomical Phenomenon – Red giant star

Nakshatra Characteristics – growing, progenitor

Collapsing molecular clouds contain primarily hydrogen and helium. Increasing core pressure yields high temperatures and eventually initiates nuclear fusion of hydrogen into helium. A star reaches stability and hydrostatic equilibrium at a few million Kelvins of energy released from fusion. As core hydrogen nears exhaustion, core temperatures continue to increase, triggering thermonuclear fusion of shell hydrogen. The star's shell begins to expand, and the star grows in size.

The inflated and tenuous outer atmosphere of a red giant star experiences a lowered surface temperature of less than 5000 K. In contrast, a blue star of Pleiades has a surface temperature of 33,000 K. The red giant Aldebaran, just 65 light years from the earth, is forty times the size of our Sun (Figure 20). Convection currents move interior matter undergoing fusion to the shell surface. Astronomers expect Aldebaran to become a white dwarf in the last stage of its evolution, expelling its shell as a planetary nebula into the surrounding interstellar medium.

Planetary nebulae provide a critical means of galactic recycling. The early universe consisted almost entirely of hydrogen and some helium. Stellar giants create carbon, nitrogen, and oxygen through nuclear fusion, interspersing them into the surrounding interstellar medium. Space scientists note unequivocally that radioactive isotopes found on Earth were freely available in the early solar system and suspect they were remnants from the explosion of a giant star from the past (Zinner, 2003). A star in close vicinity to the Earth, gradually expanding itself, preparing to leave behind its content for future generations of stars and planets reminded Rishis of the role of *Prajapati*. Aldebaran is represented by *Rohini*, the Nakshatra of *Prajapati*. Figure 20 shows giant stars evolving off the main sequence.

Orion Nebula

Astronomical Phenomenon – Giant molecular clouds
Nakshatra Characteristics – coolness, rejuvenating, conspicuousness

The Orion complex in the sky includes multiple easily identifiable stars in a vast and active region of stellar formation 1300 light years from Earth. Heightened activity in the region has created the various striking structures: the

Orion Nebula (M42), The Horsehead Nebula (IC 434), Barnard's Loop (Sh 2-276), and the Flame Nebula (NGC 2024). The Spitzer Space Telescope's infrared map of the region reveals over two thousand newly forming stars with associated disks of dust (Clavin, 2006). These disks indicate the potentiality of brand new planetary systems

The Orion Nebula, shown in Figure 22, is one of the brightest, most readily recognized configurations of stars in the night sky. Originating as a gravitationally bound collection of cold, neutral hydrogen, intermixed with traces of other elements, the nebula now consists of an ionized region around a central star emitting ultraviolet radiation. It is surrounded by an irregular, concave bay of more neutral, high-density clouds.

An abundance of cold gas clouds, ionized hydrogen atoms facilitating the formation of heavier molecules, and turbulent fronts arriving from neighboring new stars has created an ideal environment for clouds to initiate collapse. Cold expanses of star material are gently nudged together to precipitate a new creation cycle; *Soma* is synonymous with coolness and rejuvenation. The ancient *Rishis* denoted the region with the word *Mrigashira* – the head of a herd, capturing the herding of the huge population of proto-stellar systems by a bright central star ionizing the entire area.

Betelgeuse

Astronomical Phenomenon – Red Supergiant
Nakshatra Characteristics – turbulence, moisture

Having originated in the Orion cloud complex, Betelgeuse currently howls through the surrounding interstellar medium at a supersonic speed of 30 km/s. Its movement creates a bow shock four light-years wide. Among the most luminous observable stars, Betelgeuse is a red super giant in a late stage of stellar life.

Having moved off the main sequence upon exhausting its core hydrogen fusion, the star has swelled to become a red supergiant. Its contracted core now fuses helium into carbon and oxygen with the associated radiation expanding its outer envelopes of hydrogen and helium. In the future, the star will fuse the heavier elements of neon, magnesium, sodium, and silicon until it creates iron.

A thousand times the size of our Sun, Betelgeuse possesses a complex circum-stellar environment. Light from its core takes three light years to come out to the surface. The star has an average density of less than 12 parts-per-billion than that of our Sun, essentially functioning as a hot vacuum. Betelgeuse spews matter into a large plume of gas extending to a distance at least six times its stellar radius. The motion of gas in this extended atmosphere, containing water vapor, carbon monoxide, silicon, and aluminum oxides, creates bubbles as large as the star itself.

The shock waves created by Betelgeuse disperse into space as turbulence, contributing to the formation of new stars. The European Space Agency's Herschel Space Observatory confirms the existence of dense filaments of gas in its immediate environment, the width of each filament being approximately 20,000 times the distance between the Earth and Sun. Stars may form within the confines of these filaments, like beads on a string. The *Rishis* named the region for the benevolent aspects of *Rudra*, as a supporter of new beginnings.

Figure 23 shows Betelgeuse as the dark circle at the center with the complex atmosphere surrounding it.

Abel 21

Astronomical Phenomenon – Planetary Nebula
Nakshatra Characteristics – Regeneration, incubation, residency

Visually arresting emission nebulae are relatively short-lived phenomenon, lasting just tens of thousands of years, compared to the typical stellar lifetime of several billion years. Also called planetary nebulae, they form at the end of a red giant's life, the luminous core exposed by the expulsion of outer layers of complex molecules through pulsations and strong stellar winds. Intense ultraviolet radiation from the exposed core ionize and illuminate the ejected outer layer of matter. This radiant shell begins to glow as a planetary nebula. The myriad colors of these structures result directly from the rich variety of ejected molecules.

Planetary nebulae play a crucial role in the chemical evolution of a galaxy, serving as a means for a star to return enriched material to the surrounding interstellar medium. These include carbon, nitrogen, oxygen, and calcium, elements critical to life as we know it. Billions of past generations of stars have thus enriched the raw material involved in the formation of planetary systems capable of supporting life forms.

The structure of a planetary nebula is largely dictated by the magnetic fields left by its parent star. Abell 21, estimated to be over 4 light years in width, (Figure 24), is an ancient planetary nebula some 1500 light-years away in the constellation Gemini. The regeneration principle hinted at by the Sanskrit word *Punarvasu* is at play in the cycle of enriched raw material returning to the interstellar medium and the subsequent formation of new stars. Abel 21, a loosely defined container of enriched star material, is a reminder of the incubation impulse the Rishis attributed to the Devata *Aditi* - the mother of the *Adityas* (Sun).

Cancri 55's Jupiter size planets

Astronomical Phenomenon – Collapse to t-tauri stage
Nakshatra Characteristics – nourishment, expansiveness, protecting growth

By a fortuitous accident, scientists witnessed the actual formation of a new star in the R Corona Australis star-forming region. X-ray radiation from its cold expanses provides proof that matter falls towards its core at ten times the rate that can be explained by gravity alone (Kenji Hamaguchi, 2005).

An energetic process -- likely related to magnetic fields – is postulated to superheat the surface of a collapsing cloud. This brings the core to a critical stage of pre-stellar evolution, characterized by violent surface activity and strong proto-stellar winds. The central temperature during this phase[66] is not high enough to trigger thermonuclear fusion; magnetic fields appear to be the driving force.

Brown dwarfs and Jupiter-size stars are created from collapsing cores lacking sufficient gravity to ignite thermonuclear fusion. Cores with a sufficient mass, termed T-Tauri, contract further under gravitational forces to initiate nuclear

[66] Adiabatic phase or phase 0 is detectable only in near and far infrared frequencies

fusion. The still mysterious energetic process alluded to in the previous paragraph appears to assist growth towards T-Tauri mass in the core.

Brihaspati is a common reference to the planet Jupiter in Indian astronomy. Jupiter-sized bodies play an important role in the formation of terrestrial planets in a newly forming stellar system. Astronomers use the Solar Nebular Disk Model (SNDM) to explain the formation of planets around a new star. Material from the proto-planetary disk surrounding young stars initially continues to feed the central star. As the system cools, small dust grains composed of rock and ice begin to form and eventually merge into larger planetisimals from which planetary embryos form.

Planetary embryos closer to the system core become terrestrial planets while violent mergers of ice particles create distant planetary embryos. Distant proto-planets rapidly accumulate hydrogen and helium from the gaseous disk. Jupiter possesses the same ratio of hydrogen to helium as seen in the early universe. During this formation process, the gas planets lying beyond the frost line[67], cast a protecting influence over terrestrial planets. In capturing distant, often high-velocity objects from the outer reaches of the star system through their heightened masses, they prevent many of them from interacting with or destroying the delicately developing environments of Earth-like planets.

Modern astronomy has identified planetary systems around numerous stars within the Milky Way galaxy, with the term exoplanet used to refer to any planet outside our own solar system. 55 Cancri is a planetary system with five such exoplanets orbiting its primary star, 55 Cancri A. Four are Jupiter sized, and one is a large terrestrial planet. This solar system is only 40 light years from our own and has at its center a yellow dwarf star similar to our Sun. This system is a good representation of *Brihaspati/Pushya*'s characteristics of nourishment and growth. Figure 25 shows an artist's visualization of 55 Cancri E.

Epsilon Hydrae

Astronomical phenomenon – stabilization of star/planetary orbits, centripetal force

Nakshatra Characteristics – entwine, drag, curvilinear movements

---

[67] A frost line or an ice line separates gaseous planets from terrestrial type planets

Conservation of angular momentum is absolute in nature. The Sun spins due to the angular momentum it acquired from the interstellar cloud predating its formation. Even after a star has formed, the remaining material of its parent nebula continues to rotate due to left over angular momentum. Its rotation forces material to flatten into a disc that rotates around the new star, termed a proto-planetary disc. Conservation of angular momentum dictates how far and fast planets will orbit from their respective suns as well as how fast they will spin around their axes. The orbital and rotational characteristics of planets -- Earth's 24 hour day, Saturn's 10,747 day-long year – are born out of the fundamental formation of planetisimals.

Physics dictates that an object experiences acceleration while in motion along a curved path. A complementary force is needed to balance the system, whether friction for a car rounding the bend or tension for a ball swung on a string. In a stellar system, the central star provides this force in the form of gravity. A planet is forced to migrate to a newer orbit if its corresponding gravitational pull does not balance its motion. Migration of proto-planets is quite common in nascent star systems, with the gas and the debris present in the proto-planetary disc introducing a drag effect on migrating planets. This drag can counter any excessive migration.

Epsilon Hydrae is a simple star system. Located 130 light-years from Earth, it consists of a binary star system and two other components. Its binary star pair, ε Hya AB, has an orbital period of 15 years. Its other star pair, ε Hya C, is a spectroscopic binary system with a period of 9.9047 days. This pair also orbits ε Hya AB with a period of 870 years. The last star in the system, ε Hya D, has an orbital period around the AB pair of 10,000 years. The paths traced by these stars can be visualized as the slithering movement associated with a *Sarpa* (snake), the Devata of *Ashlesha* Nakshatra. Figure 26 shows the phases in the formation of planets. Excess disc material can be seen in earlier phases when proto planets are settling into their steady state orbits. Figure 26 is an artist's visualization of the different stages in the formation of planets around a new star.

Leo I dwarf galaxy

Astronomical phenomenon – Satellite galaxies, dwarf spheroidal galaxies
Nakshatra Characteristics – Heritage, defunct, redundant

A dwarf spheroidal galaxy is generally a low luminosity companion of a larger galaxy. Unlike a dwarf elliptical galaxy, a dwarf spheroidal galaxy contains minimal gas, dust, and recent star formation. By studying the motion of the galaxy's component stars, astronomers note that dwarf ellipticals must contain many times more mass than is visually apparent. Cold Dark Matter cosmologists propose the missing mass is attributable to dark matter within the galaxy.

Leo I, a dwarf spheroidal galaxy adjacent to the region of the star Regulus in the night sky, is one of the most distant satellites of the Milky Way galaxy at 820,000 light years from Earth (Figure 27). Hubble Space Telescope observations confirm the galaxy went through a sudden growth spurt a few billion years ago, but about half a billion years ago, star formation ceased. Scientists have also confirmed that this galaxy does not rotate. Leo I dwarf seems to be embedded in ionized gas equal in mass to its visible portion. The defunct, static nature of Leo I and the static nature assigned to ancestors in the Vedas are similar. The *Rishis* assigned the Pitru Devatas, whose functions relate to ancestry, to this Nakshatra.

The galaxies of the universe began forming 13 billion years ago; a new study shows that the Milky Way formed along with old globular star clusters and dwarf galaxies, some of which remain intact within the inner halo with the remainder merging to form the current structure of the galaxy (Waller, 2013). Newer galaxies which formed in the vicinity of the Milky Way also merged into the larger structure, their stars creating the outer halo. Thus the current composition of the Milky Way is attributable to an ancient heritage of multiple dwarf galaxies.

The Sagittarius dwarf galaxy is currently merging into the Milky Way, demonstrated by volumes of data from 2MASS, an infrared sky-survey project led by the University of Massachusetts. The Sun may be a part of this dwarf galaxy.

NGC 3842

Astronomical phenomenon – Clearing of the proto planetary disc, plasma wind

Nakshatra Characteristics – piercing, vigorous, teeming with, day-night cycle

A planet forms from the disc of gases and particles swirling about a new star; as a proto-planet, it accretes mass from its vicinity until the excess material of the disc is exhausted. Planets settle into stable orbits as debris is removed from the disc, and, in turn, diurnal cycles on planets are established when planetary rotations settle.

Vedic texts attribute day-night cycles to *Aryama*, the Devata of the Nakshatra *Falguni*, praised for his prowess to clear. Astronomers note two mechanisms that rid a proto-planetary disc of excess debris: the powerful stellar wind of a newly formed star and the heating from its electromagnetic radiation. Debris material, excited by radiation, accelerates towards the edge of the newly formed solar system. The intensity of this radiation is larger at the onset of fusion and is capable of clearing excess gas.

A disc also contains solid dust particles which tend to linger longer than their gaseous counterparts. Solar wind from a young star forces away particles lying just outside the central star's gravitational radius with particles inside this radius cleared by yet another mechanism: the polar jet. Class O and B stars in the neighborhood may also assist in the clearing process.

Solar winds are scaled versions of galactic winds, streams of high speed charged particles emanating from the center of a galaxy. Galactic winds push gases from near the core into the intergalactic medium and into the halo of the galaxy while quasars generate winds of the highest strength. The Leo galaxy cluster contains NGC 3842, a large elliptical galaxy surrounded at its center by three quasars, shown in Figure 28 . *Falguni* Nakshatra is the region of sky with the ecliptic at its highest tilt with respect to the galactic plane. Distant galaxies are clearly visible here without obstruction by the dust lanes of the galactic disc.

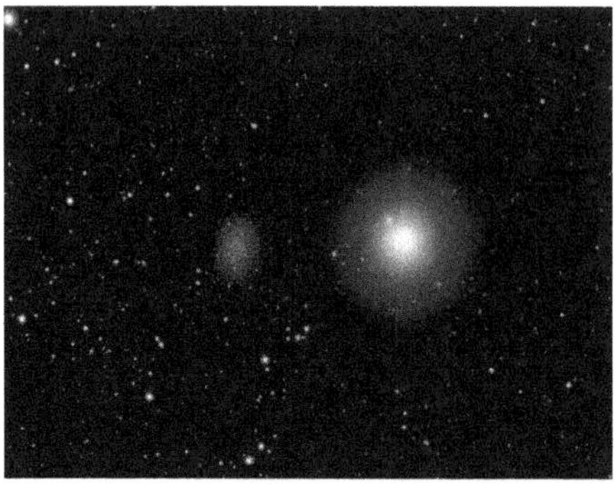

Figure 27 : Leo I satellite galaxy to the left of Regulus – Picture Credit NASA/Hubble

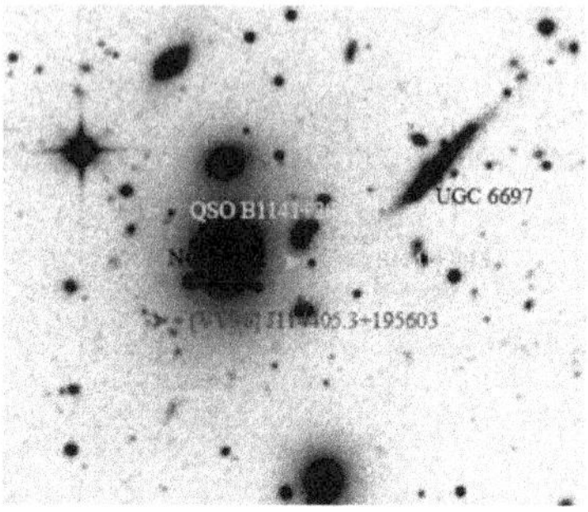

Figure 28 : Leo Galaxy Cluster, Picture Credit – Paul Beskeen

Figure 29 : Whirlpool galaxy – Picture Credit: NASA/ESA

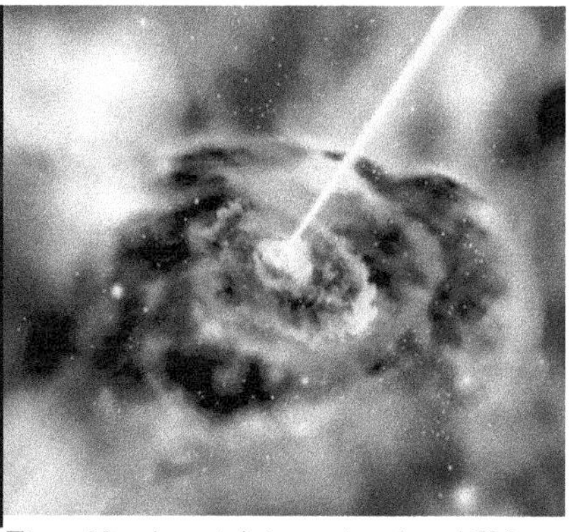

Figure 30 : An artist's impression of an AGN. Photo credit: ESO website.

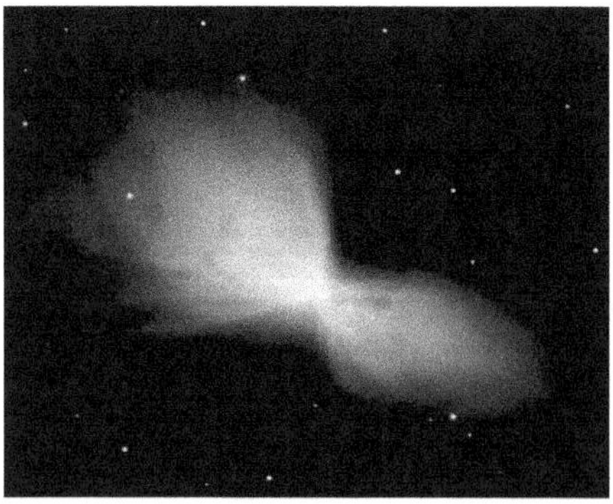

Figure 31 : Bipolar jet from a young star. Image credit: NASA, STScI.

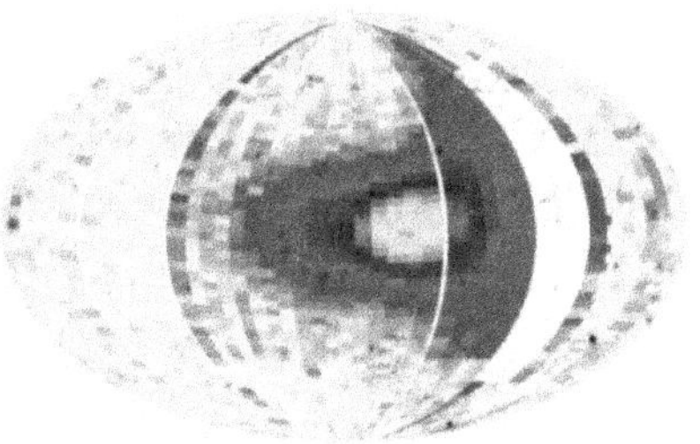

Figure 32 : Intensity of material arriving from outside our solar system. The maximum intensity in in the circular spot. Map credit NASA's IBEX spacecraft.

Whirlpool galaxy

Astronomical phenomenon – Star bursts, ionized HII regions
Nakshatra Characteristics – brightness, creative impulse, bountiful

Collisions among galaxies are common and represent an important means of galactic evolution. Unlike conventional collisions between rigid bodies, galactic collisions are gravitational interactions with exceedingly minimal probabilities of individual stellar collision. A larger galaxy may, of course, irrevocably transform the structure of a smaller participant during a merger. In some cases, a pair of galaxies may pass through each other multiple times before merging during the final pass. Minor gravitational interactions between a pair of galaxies are also possible when they move within one another's vicinities, rather than through each other. An example of this type of interaction is the skew seen in a primary galaxy's spiral arms when a satellite galaxy moves close to it.

Besides changing the appearance of a galaxy, gravitational interactions can cause subtler changes like density waves sweeping through the hydrogen clouds within interacting galaxies. Huge clouds of hydrogen, either in atomic or molecular form, lie scattered throughout galaxies. Gravitational interaction produces compression waves, forcing previously diffuse areas of gas into tighter pockets of opaque and dense gas. A highly compressed region can collapse under its own gravitational pull, giving birth to a star at the center. When this happens, a new-born star heats large regions of space, and stellar winds finally sweep away the surrounding layers of dust and gas.

As hydrogen clouds are so massive, a star is seldom created in isolation; instead, many are born in rapid bursts. The Vedic concept of *Bhaga* is associated with creative impulse and is honored as the provider of all things auspicious or life supporting.

Star burst regions, even at several million light years distance, are noticeable in the visible spectrum. The Whirlpool Galaxy, shown in Figure 29 , is a grand-design spiral galaxy, and is one of the best known galaxies in the sky. Easily observed by amateur astronomers, the galaxy and its companion, NGC 5195, are commonly analyzed to further understanding of galaxy structure and interactions. The pronounced spiral structure of the Whirlpool Galaxy is believed to be the result of the close interaction between it and its companion.

Recent simulations studies confirm that the spiral structure of the Whirlpool Galaxy was caused by NGC 5195 passing through the main disc 500 to 600 million years ago. The smaller companion lies slightly behind the Whirlpool galaxy today as can be seen in Figure 29. The Vedic Rishis employed the word *Falguni* to characterize this region, an apt characterization given the high intensity of creative impulse in its star burst zones.

NGC 4552

Astronomical phenomenon – Bright, obscured objects, obscure AGNs
Nakshatra Characteristics – reveal, bright, assist comprehension

Active Galactic Nuclei (AGN) such as quasars, blazars, and Seyfert galaxies are the most luminous of objects in our Universe. The energy from one AGN is the equivalent of that emitted simultaneously from billions of stars in a compact space. The Swift and Suzaku telescope teams recently, discovered a class of AGN that constitutes 20% of the population of AGNs in the local universe (Winter, 2008). Shrouded in deeply obscuring dust and gas, AGNs are detectable only through high-energy X-ray emission.

The more abundant variety of AGN is partially covered by a torus-shaped envelope of materials, but a more recently discovered type of AGN is surrounded completely by an obscuring shell. A bright source of concealed light is evocative of the nature of the *Nakshatra Hasta.* NGC 4552 is the first galaxy discovered with an envelope. Figure 30 shows an artist's visualization of an AGN.

Seyfert galaxies

Astronomical phenomenon – Prevention of excessive spin, Clear accretion disc**,** Polar Jets
Nakshatra Characteristics – sharp, distinct, chiseled, beautiful

An accretion disc surrounds a newly formed proto star. Previous sections have described how solar winds and radiation from a young star clear excess material from a proto stellar disc, but the material closest to the center remains stationary even after the completion of this process. Gravitationally bound by the newly formed proto-star, material in the disc closest to the center feeds into stellar accretion and increases the star's spin rate. The newly formed stellar system becomes stable when all excess material is cleared.

Polar jets provide a final opportunity for excess material to escape the gravitational pull of a new star. Continuous energetic flow from its poles relieves the accumulation of angular momentum from the accretion disc. Bipolar outflow is an important determining process in final stellar shape and structure.

Twashtra is praised in the Vedas as the cosmic architect who chisels fine shapes in objects, bringing them beauty; the meaning of the Sanskrit word *Chitra* is "pleasant to see." According to a narrative from Sanskrit texts, the effect of the influence of *Twashtra* is to cause the Sun to shed excess brilliance.

Mass ejected by a young star in powerful polar jets creates a supersonic shock front. Astronomers detect this shock front in the infrared spectrum from the heated gas in the vicinity of the jets. Called a molecular bow shock, only the youngest stars possess them. Jets from more evolved young stars -- T Tauri stars -- produce similar bow shocks visible in optical wavelengths. Figure 31 shows polar jets from a young star.

Accretion disks and the material clearing phenomenon associated with them are associated with both young stars as well as AGNs, and bipolar jets are common to both. Jets from AGNs make those of proto-stars appear mild in comparison; the most powerful of these are termed relativistic jets and carry plasma churned out from the centers of galaxies and quasars. Relativistic jets reach several thousand light years in distance. Seyfert galaxies- Mrk[68] 69, 268, 270 and 279 - in the Chitra Nakshatra area of the sky evoke again the role of *Twashtra.*

Neutralized galactic wind

Astronomical phenomenon – Galactic wind
Nakshtra Characteristics – movement, fierce, strong, wind, storm

Our solar system sits inside a giant magnetic bubble filled with charged solar particles. The edge of the bubble is called the heliosheath, which creates a natural boundary between the region of charged particles and the rest of the

[68] Mrk : Nomenclature of sky objects identified by Russian astronomer Markarian

galaxy. The heliosheath is in turn elongated as our solar system travels through space at high speed and is compressed by galactic wind.

Galactic winds carry highly charged particles that strike the heliosheath and rebound through the strong magnetism that retains the shape of the bubble. As a result, the heliosheath prevents harmful, highly energetic, charged particles from entering the solar system. Since magnetism does not influence neutral particles, uncharged matter may pass through. Thus, neutral atoms from the Sun's galactic neighborhood progress steadily towards the planets at high speeds, traveling inwards and deflected by the Sun's. In the Vedas, *Vayu*'s energy moves the winds and controls meteorological events on Earth; a galactic extension of this principle led the Rishis to assign Vayu to this region in space.

The IBEX (Interstellar Boundary Explorer) satellite, designed to study the interstellar region around the Sun, has sampled multiple heavier elements from the Local Interstellar Cloud and has also measured the strength and direction of interstellar wind. A full-sky neutral atom map produced by IBEX shows the strongest winds to lie in the Libra region of the sky. Scientists, however, identify the origin of strongest interstellar winds in the direction of Scorpio. IBEX satellite makes its observations in Earth's orbit. This discrepancy (shown by the circular spot in Figure 32) is explained by the Sun's gravitational deflection of the wind of neutral atoms.

The map shown in Figure 32 was generated from NASA's IBEX and maps the material flux arriving from outside our solar system.

Far end of galactic central bar

Astronomical phenomenon – Raw stellar material branching into spiral arms
Nakshtra Characteristics – branching, fierce energy, central control

The star population of the Milky Way Galaxy is unevenly distributed; more stars lie along the spiral arms than outside of them. The galactic bulge contains the highest stellar density region within the Milky Way, with approximately 10 million stars known to orbit within a single light-year of the galactic center. In contrast, the 4.2 light year distance between the Sun and our nearest star is indicative of the sparseness of stars along spiral arms. The Milky Way's central bulge influences not only stellar birth and death in the core but also controls overall galactic evolution.

Recent infrared observations by the Spitzer Space Telescope have clarified the number of major and minor arms of the Milky Way. Six spiral arms branch out of the central bulge, of which two are "major": the Scutum-Centaurus and Perseus. Astronomers demoted two arms, the Norma and Sagittarius, to "minor" status, and, having studied galactic gas, classified the last two arms as "minor enclosing." At approximately 3 kilo parsecs from the galactic center, the latter are termed "near" and "far" 3kps arms. Major arms possess high stellar densities while minor arms have lower densities and contain both young and old stars with abundant pockets of star-forming activity. An artist's visualization of the structure of the Milky Way galaxy, with its associated central bar, is shown in Figure 33.

Three arms fork from each edge of the central bar of the galactic bulge; the bar itself contains a varied distribution of clumps of giant stars. The discovery of these clumps, confirmed through DIRBE infrared maps, aid in NASA's estimate of the size of the bar through the Spitzer Space Telescope survey of some 30 million stars within the bulge. Astronomers estimate the bar to be 27,000 light-years across, at a 45 degree angle from our Sun to the galactic center. The location of the *Vishakha* Nakshatra in the sky lines up almost exactly in the direction that the Milky Way galaxy's far arms branch out from the central bar, and may be no accident.

Galactic black hole

Astronomical phenomenon – Black Hole at the center of the Milky Way galaxy

Nakshatra Characteristics – root, source, gulp, destroy

Supermassive Black holes reside at the center of elliptical galaxies; the gravity of a black hole is extreme, with contained matter being highly compressed. Black holes are so named due to the inability of light or radiation to escape their confines. In his general theory of relativity, Albert Einstein predicted that a sufficiently compressed mass can theoretically deform space-time; black holes serve as evidence.

As various conventional laws of physics appear potentially violated within a black hole, physicists define a black hole's event horizon as the boundary outside which the laws of physics hold true. Matter or light entering an event horizon simply disappear. Due to its characteristic invisibility, the presence of a black hole can only be inferred through the behavior of matter and electromagnetic radiation in its vicinity.

A black hole spins around itself, a disc of material collecting just outside its event horizon. Matter falling in from this accretion disc shines bright before disappearing, and thus, the neighborhood of a large black hole ranks among the brightest spots in the universe.

The central region of the Milky Way falls into six Nakshatras- *Vishakha*, *Anuradha*, *Jyeshta*, *Mula* and the *Ashada* pair. These Nakshatras, with their associated Devatas, capture the ideas behind the various forces in play in the densest regions of our galaxy. Three galactic arms merge at the far end of the central bar in the *Vishakha* while the three other arms fork out from the near end of the central bar in the region of *Ashada (Uttara)*. In Vedic terminology, *Vishve Devas*, the Devatas associated with *Ashada (Uttara)*, are associated with the limbs. The word *Mula*, defined as "root", is thus an appropriate assignment for the region hosting the galaxy's central black hole. The fierce characterization of the Nakshatra *Mula*'s Devata *Nirriti* as one who devours creation is an apt symbolism for the central black hole. Figure 35 shows gas streaming out of the accretion disc, escaping away from the central black hole.

Density waves

Astronomical phenomenon – Propeller action of the central bar of our galaxy

Nakshtra Characteristics – central control, a powerful force, prominent, invincible

Astrophysicists point to the important role of the Milky Way's central rotating bar in sustaining the continued creation of new stars in our galaxy. The bar's rotation creates spiral density waves through a resonance excitation mechanism. These waves transport angular momentum between the bar and the spiral arms, while matter in the spiral arms moves either inward or outward depending on the direction of angular momentum. Matter flowing inwards carries molecular hydrogen that fuels the black hole; matter flowing outwards initiates starburst activities in the arms.

In Vedic terminology, *Indra* is responsible for the main control force of complex systems. *Indra* supports the tendencies of expansion and overpowers the tendency of contraction associated with *Vritra*, his arch enemy. No other area of the sky can represent this aspect of Indra. Figure 34 illustrates various density waves, and, as shown, the Milky Way is a barred spiral galaxy with a central bar whose strong propeller action maintains the density wave (d).

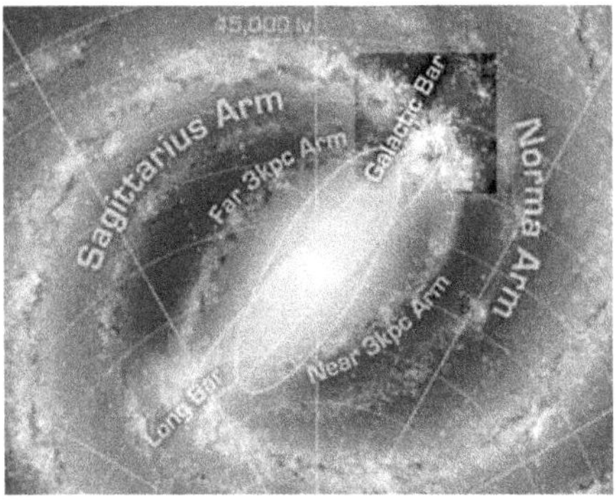

Figure 33 : Artist's visualization of the structure of the Milky Way galaxy, showing the central bar. Picture Credit : NASA

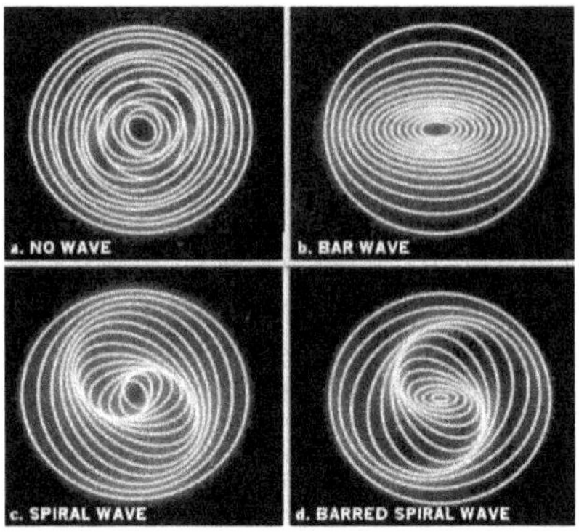

Figure 34 : Density waves. Picture credit – Universe review

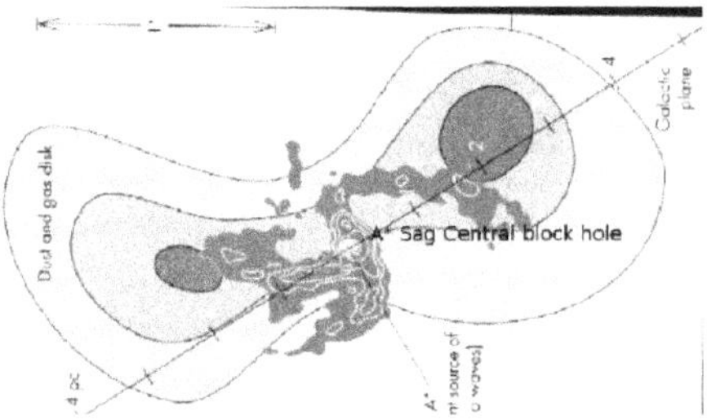

Figure 35 : Gas streaming out of the accretion disc around the central black hole, escaping away from the central black hole. Picture credit – University of Oregon

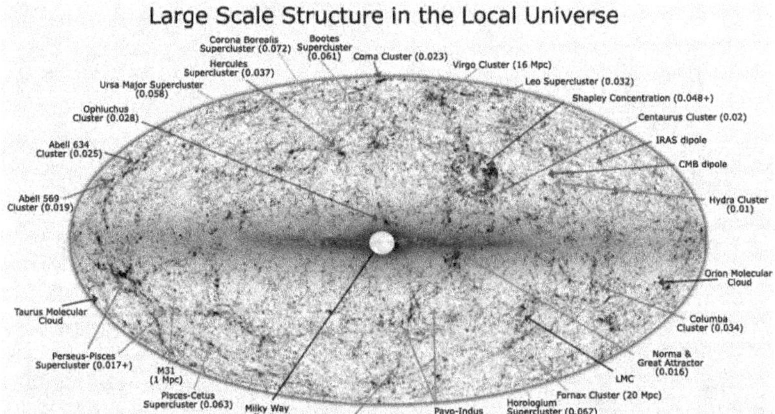

Figure 36 : Image of the cluster of galaxies produced by UMASS-2 scan of the sky. Picture Credit – T. Jarrett IPAC/Caltech

Shapely Supercluster

Astronomical phenomenon – Complex gravity interactions in tightly packed systems

Nakshtra Characteristics – friendship, affinity, order, stability

The Milky Way and its neighboring galaxies are currently moving towards the mysterious "Great Attractor," located 250 million light years away. Recent findings suggest that a large conglomerate of galaxies beyond the Attractor is exerting gravitational pull on our galaxy in its general direction. The Great Attractor's direction in space is known for a preponderance of large, old galaxies, many of which are colliding with their neighbors.

The Great Attractor lies in the direction of the Shapley Supercluster, 500 million light years away. The Shapely cluster, shown in Figure 36 is the most massive among known superclusters. Light originating from a star in our galaxy's central bulge reaches a neighboring star as quickly as a week due to the high density of stars in this region. The gravity forces here act in delicate balance, holding systems of stars in relative stability. The *Anuradha* Nakshatra and its Devata *Mitra,* denoting the principle of affinity, were aptly assigned to this region of the sky by the Rishis.

RSCG1-3 & Alicante 8

Astronomical phenomenon – Enriching of galactic raw material through red supergiant-supernova cycles

Nakshtra Characteristics – invincible, pliable, dynamic, flowing, life supporting

Four groups of massive young stars -- RSCG1, RSCG2, RSCG3 and Alicante-8 -- were discovered through the use of infrared scans of the region adjacent to the intersection of the near end of the central bar of the Milky Way and the inner section of the Scutum–Centaurus Arm. These clusters of red giants are obscured in visible light frequencies, while the weight of each is between 20,000 and 50,000 solar masses. The stars in these groups are massive as well as young -- less than 15 million years old, while the average star in our galaxy is between

7 and 12 billion years old. Stars in RSCG-1, 2, and 3 are type-II supernova progenitors.

A Type II supernova is created from the rapid collapse of a massive star, eight times the mass of our Sun. Fusion progresses rapidly in a supernova progenitor star towards heavier elements -- hydrogen, helium, carbon, neon, oxygen, silicon, nickel and iron. The fusion of iron or nickel does not produce net positive energy and thus causes the star to collapse rapidly in the initiation of a supernova, which, in turn, spews rich elements back into the galaxy as raw material for future stars.

The Sanskrit word *Apah* commonly refers to water, a fundamental life sustaining material. Vedic Rishis looking for an extension of this life sustaining principle in space found it here and named it the Devata *Divya Apah*. Figure 37 depicts an artist's conception of the interior of a red supergiant ready to go supernova.

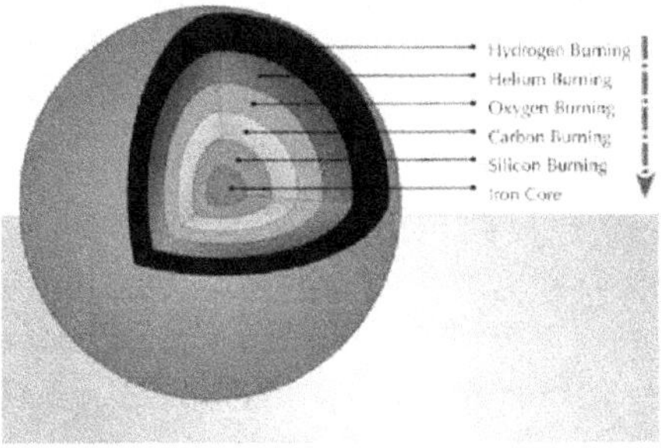

Figure 37 : Inside of a red supergiant that is ready to burst out as a supernova – image credit science in schools

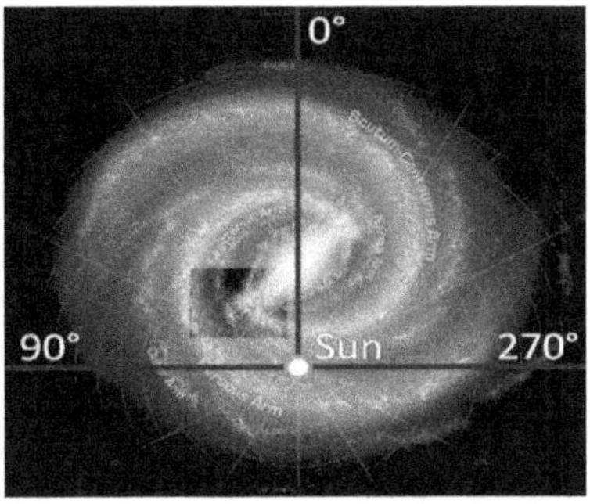

Figure 38 : Elongated central bar of the Milky Way galaxy showing the arms branching out from its near end – picture credit NASA

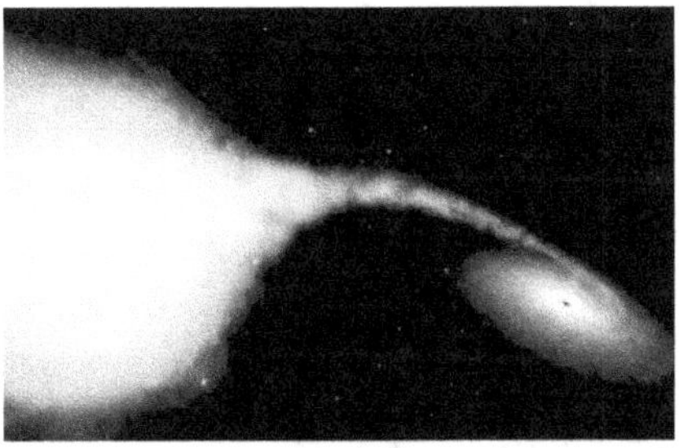

Figure 39 : Solar wind from the companion star feeding the accretion disc of the black hole in Cygnus X-1. Picture Credit - ESA

Figure 40 : Cluster of stars that form M15. Picture Credit – Hubble Space Telescope

Figure 41 : NGC 7331 Picture Credit – Vicent Peris using the 3.5 meter telescope at the Calar Alto Observatory in southern Spain

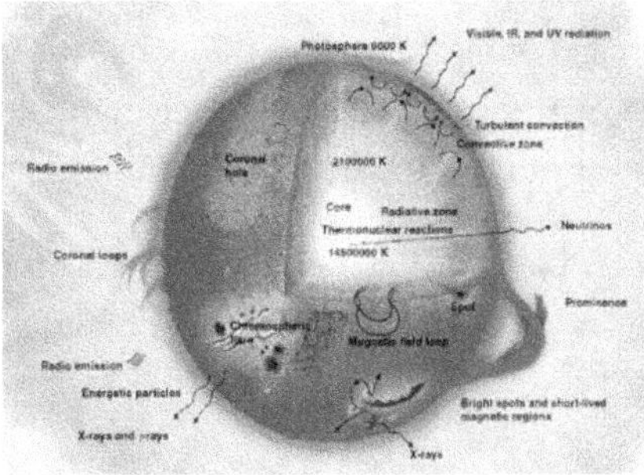

Figure 42 : Alpha Andromeda Picture Credit Imagine the Universe web site, High Energy Astrophysics Science Archive Research Center, NASA Goddard Space Flight Center

Near end of galactic central bar

Astronomical phenomenon –Ionized hydrogen streaming out to the star creation regions

Nakshtra Characteristics – invincible, funneling, encapsulating

The near end of the Milky Way's central bar lies in the direction of *Ashada (Uttara)* Nakshatra, symmetrical to its far end located in the *Vishakha* Nakshatra. The raw material used in the formation of new stars in the arms is energized within the central bar and pushed out through the bar's end points into the galactic arms. Powerful density waves move star raw material from the near end of the central bar into the Scutum-Centaurus arm, the Norma arm and the Near 3 kpc arm.

The funneling of energetic matter, initiating star bursts elsewhere in the galactic arms, is a reminder of the function of *Vishwedevas* on a galactic scale. Figure 38 shows the elongated central bar of the Milky Way galaxy and the arms branching out from its near end.

Cygnus X-1

Astronomical phenomenon – Stellar black hole

Nakshtra Characteristics – swift, far reaching, hidden, auditory, solid

A *stellar black hole* is formed by the gravitational collapse of a massive star, inevitable following the exhaustion of sources of fusion. The collapsing mass of the star becomes super compact, creating a white dwarf or a neutron star, or, if the catastrophic gravitational limit is breached, a black hole.

A black hole retains the angular momentum of its progenitor, while matter in its accretion disk similarly maintains spin. Heat and other radiation may be produced due to the friction caused by differential spin of matter in the accretion disc.

Cygnus X-1 was the first X-ray source in the sky to be accepted as a black hole and is one of the most studied astronomical objects of its class. Extremely compact, it is fourteen times the mass of our Sun but packed within a radius of just 25 km. Located about 6000 light years from the earth, this black hole has a companion, the blue supergiant HDE 226868. The binary pair orbit one another every six days. HDE 226868 continually sheds its mass as stellar wind, losing the equivalent of the mass of our Sun in 400,000 years to its companion, Cygnus X-1.

The stellar wind from the blue supergiant supplies material for the accretion disk around the black hole; this rapidly rotating disc heats matter to millions of degrees, generating flares of hard X-rays. The accretion disk also sheds matter through polar jets associated with the black hole. High energy gamma rays[69] are created where the relativistic jets meet solar winds from the companion blue supergiant.

The solidity attributed to *Sharavana* coincides with the compactness of the black hole, while the far reaching radiation of Cygnus X-1 reminds one of the swiftness of *Vishnu.* High energy Gamma rays and hard X-rays penetrate and spread out to the edge of the universe, while their source lies hidden from view like the feet of *Vishnu* obscured in dust. Figure 39 illustrates the solar wind from the companion star feeding the accretion disc of the black hole in Cygnus X-1.

M15

Astronomical phenomenon – Globular clusters in our galaxy's halo
Nakshatra Characteristics – residing, flow steadily, age-old

A globular cluster is a spherical collection of stars that orbits a galactic core as its satellite. Stars in a globular cluster are bound by gravity; globular clusters are spherical with higher density toward their centers. Their brightness and appearance make them distinct from younger open clusters. Stars within a globular cluster, compared to the general population of stars in the galaxy, are poorer in elements other than hydrogen and helium. Astronomers classify the population of globular clusters as those found in the galactic halo, away from the thin disk of the galaxy and those strongly concentrated toward the center of the galaxy, at distances less than that of the Sun from the galactic disk

A halo cluster is often found at a distance of ten kilo parsecs or more, above or below the galactic plane. It orbits around the galactic center, approximately

[69] Above 100 GeV

in the general rotation direction of the galaxy though some halo clusters move opposed to galactic rotation.

Globular clusters contain many of the oldest stars in the galaxy, with astronomers having identified 140 such groupings in the Milky Way. Globular clusters do not form new stars. Near collisions and close interactions among resident stars are common due to high stellar density, with such interactions capable of producing exotic classes of stars: blue stragglers, millisecond pulsars, and low-mass X-ray binaries.

M15 is a globular cluster orbiting the galactic center at over 30,000 light-years from the Earth. Its 175 light yearlong structure has the combined luminosity of 360,000 Suns, and the cluster is among the densest at over 100,000 stars. The stars close to its center likely revolve about a central black hole. M15 contains numerous variable stars and pulsars and even a double neutron star system and a planetary nebula. Three other globular clusters -- M2, M30, and Segue – join it in the galactic halo.

These four globular clusters mirror the characteristics of the Nakshatra *Sharvishta*: ancient, moving steadily, and maintaining shape since the formation of the galaxy. The Sanskrit word *Vasu* is associated with the concept of dwelling; globular clusters remind us of the grouping of the eight Devatas, holding the cells of the physical body where consciousness dwells. Figure 40 shows the cluster of stars that form M15.

NGC 7492

Astronomical phenomenon – Primordial hydrogen, flattened globular structures

Nakshtra Characteristics – spreading, healing, occupying, expanding

The visible universe is dominated by hydrogen (73%) and helium (25%); most of everything else on Earth is made up of the remaining 2% of the visible universe. Hydrogen and helium nuclei were produced in the hot, dense conditions immediately following the Big Bang when the universe began to expand and cool. Just as steam condenses into water droplets, quarks released during the Big Bang condensed into baryonic matter -- neutrons and protons -- when the universe began to cool. The nucleus of a hydrogen atom contains only a single proton. Further expansion and cooling of the universe allowed some protons to combine with neutrons to form helium nuclei – two protons and one neutron. The expansion of the first three minutes following the Big Bang created all hydrogen and helium found in the universe today.

Through the study of extremely distant galaxies, scientists estimate the amount of baryonic matter in existence when the universe was only a few billion years old, but matter in the observable universe is only a sixth of this estimate. Numerous theories have attempted to account for this missing matter over the past decade. In a recent study, a team of scientists (A.Gupta et al, 2012), using ESA's Chandra and Japan's Suzaku satellite, found gas surrounding the Milky Way to be at temperatures hundreds of times hotter than that of the surface of the sun. Other studies have shown that the halos of the Milky Way and other galaxies contains hot gas much more massive than previously observed gas in the halos. The hot gas halo is thus, one area where missing baryonic matter appears to be hiding.

NGC 7492 is a flattened globular cluster in the outer halo, located at a distance of 110,000 light years from Earth. A relatively sparse globular cluster whose outer members have likely moved away from the cluster, NGC 7492 is populated by stars whose masses have little variation with respect to their distances from the cluster center. The Vedic concept of *Varuna* expands and occupies; *Varuna* is commonly seen as the Devata of waters as water is a representation of fluidity. Hydrogen is the most common element in the cosmos just as water is the most abundant material on Earth.

A flattened globular cluster serves as a reminder of *Varuna*'s tendency to support expansion and dissemination. The hot outer halo is the most expanded aspect of the Milky Way; Figure 43 shows the extent of the outer halo. The location of the Large and Small Magellanic Clouds are shown around the Milky Way Galaxy to provide depth context.

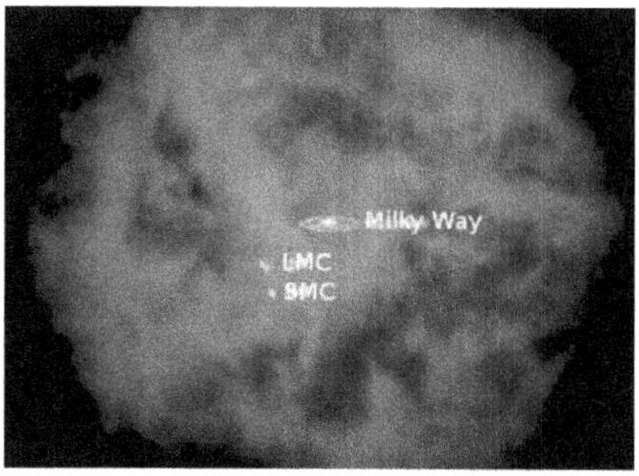

Figure 43 : Hot Halo of the Milky way with the location of Large and Small Magellanic Clouds are to provide a depth perspective. Credit: Illustration: NASA/CXC/M.Weiss; NASA/CXC/Ohio State/A Gupta et al

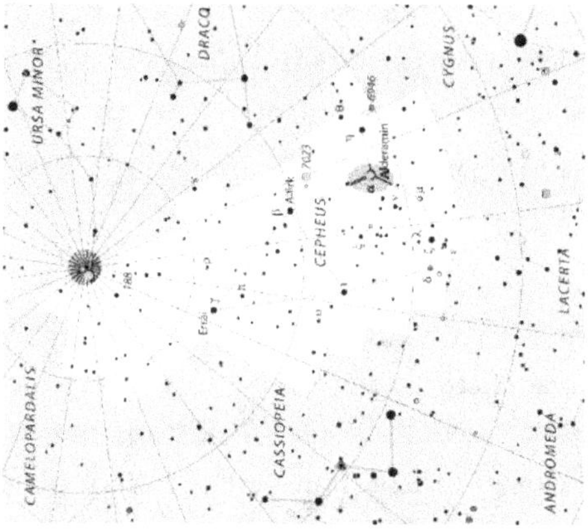

Figure 44 : Constellation Cepheus, the current pole star and star alpha Cephei close to the circumpolar ring. Picture credit IAU Sky

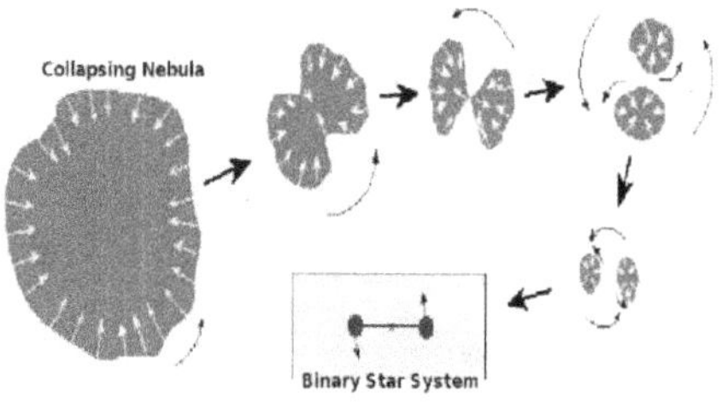

Figure 45 : A collapse of a molecular cloud splits into two with original angular momentum transferred three ways. Picture credit – UT Knoxville, Physics Department

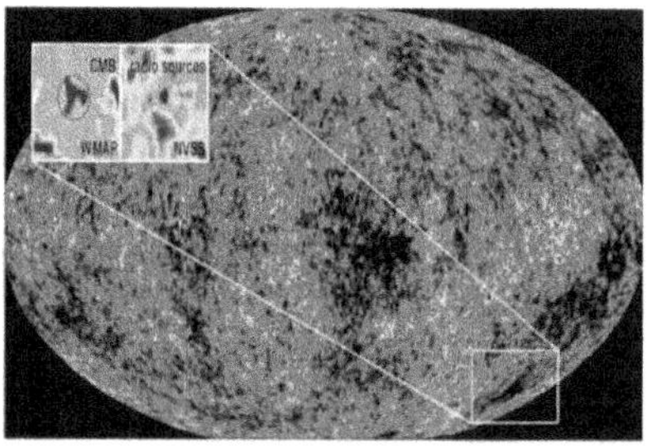

Figure 46 : Cold Spot shown in map of the CMB emissions. Image Credit : Image credit: NASA/WMAP Science Team and Rudnick et al. NRAO/AUI/NSF

NGC 7331

Astronomical phenomenon – Counter spin, memory in the cosmos

Nakshtra Characteristics – causative, contradiction of still and mobile, elevated, no beginning

One of the stranger astronomical discoveries of the past century is the presence of an excess of counter-clockwise rotating galaxies in our universe. The discovery of a "favored" spin direction among galaxies as far as 600 million light years away implies a memory of angular momentum from before the initial Big Bang. A universe starting "fresh" should not have this memory; indeed, asymmetry on such unimaginably large scales cannot statistically be expected to spring from nothingness. Skewed galactic angular momentum leads to speculation of either the presence of other universes or of causal events before the Big Bang.

Retrograde motion within galaxies is another puzzle related to angular momentum. Galactic retrograde motion is generally attributed to galactic mergers, but this explanation does not fit the case of galaxies without any other proof of past mergers such as distorted shape.

The most drastic of retrograde motion is observed in the central bulge of NGC 7331, a galaxy is similar in size and structure to the Milky Way. Often referred to as the Milky Way's "twin," it lies 40 million light years from Earth. In spiral galaxies, the central bulge typically co-rotates with the disk, but the bulge in NGC 7331 rotates in the opposite direction. One possible explanation is that the current bulge formed from in-falling external material, but failing this, other attempts at explanation are currently difficult.

The Rishis cognized the subtlety of the cosmos in this region, giving it a name which, when properly interpreted, is a contradiction. Its given name, *Aja Ekapad,* points to what escapes our logic, such as the universe remembering a preference for counter clock wise rotation for its galaxies. The retrograde motion of NGC 7331 arising without in falling material can be seen as a good representation of this contradiction.

The Milky Way's disc is visible high in the sky in *Proshtapada* Nakshatra and is, in fact, seen at its highest elevation in the night sky with respect to the ecliptic. *Proshtapada* refers to an elevated support structure. Figure 41 shows NGC 7331 as seen through high resolution telescopes.

Alpha Andromeda

Astronomical phenomenon – Convective stability in stellar atmosphere, chemically peculiar stars

Nakshtra Characteristics – fragile, a certain type of movement, atmospheric, elevated structure

Astrophysics categorizes stars with distinctly unusual metal abundances in their atmosphere and surface layers as "chemically peculiar," connoting stars originally formed from gas clouds with normal chemical compositions. Unusual metal abundances in their surface layers are thought to be caused after the stars formed, either through diffusion or magnetic effects.

The elements Manganese, Strontium, Yttrium, Zirconium, Platinum, and Mercury can rise from the interior to the surface of a star whose atmosphere is stable enough for convection. In most stars, convective mixing prevents the display of atmospheric metal abundance. Astronomers attribute convective stability to the unusually large magnetic fields generally observed in stars of this type.

Alpha Andromeda is categorized as a peculiar star; it exhibits ionized mercury lines in its spectrum and is therefore designated to be a Mercury-Manganese star. With an excess of elements -- phosphorus, manganese, gallium, strontium, yttrium, zirconium, platinum and mercury in its atmosphere – Alpha Andromeda lacks a strong dipole magnetic field. Its relatively calm atmosphere owes to its relatively low rotation rate; in a calm atmosphere, some types of atoms may sink under the force of gravity, letting radiation pressure lift others towards the exterior. Alpha Andromeda is the brightest of the Mercury-Manganese stars in the night sky.

Ahir Bhudnya's influence is seen in certain movements in the atmosphere -- those that arise from the depth and are fragile. Alpha Andromeda is a representation of this principle. Figure 42 shows an artist's visualization of the convection phenomenon.

Alpha Cephei

Astronomical phenomenon – Precession of Earth's poles

Nakshtra Characteristics – nourishment, paths, unequal length of day light, precession

The north pole of the Earth points in the direction of the star Polaris. It will gradually shift from there before returning in approximately 23,000 years. This phenomenon is called the precession of the poles and causes the seasons to shift over centuries. On an imaginary circle in the sky created by the earth's North Pole, due to precessive effects, lies the star Alpha Cephei, which will become the Pole star in 7500 CE. Alpha Cephei is currently the Pole star of the planet Mars. A white Class A star which will evolve to a red giant, Alpha Cephei, is 50 light years away from Earth and spins once in 12 hours – relatively fast compared to our Sun's spin rate of once a month.

Vedic astronomy uses the sidereal coordinate system, with longitude measured along the ecliptic instead of along the celestial. The conventional coordinate system follows an epoch based convention of fixing its zero longitude point. The J2000 coordinate system, for example, ties its zero longitude point to the position where the spring equinox occurred in 2000 CE. The sidereal system instead uses a fixed zero point that falls in the Nakshatra Revati.

Vedic astronomy has recognized precession of the equinox since antiquity, measuring the phenomenon with reference to the said zero point in *Revati*. *Pusha* is the Vedic concept that explains the rise of solar days and nights and is the master of paths. This ostensibly cryptic idea points to Pusha's association with precession. Alpha Cephei, a second magnitude circumpolar star is a good representation of *Revati* Nakshatra and its Devata *Pusha*. Figure 44 shows the constellation Cephei, the current pole star, and the star Alpha Cephei close to the circumpolar ring.

Beta Arietis

Astronomical phenomenon – Systems of stars, binary stars
Nakshatra Characteristics – messengers, healers, swift moving, yoke, easy to harness, sensitive

A third of the stars in the Milky Way are estimated to be binary or multiple. Binaries have an edge during the star formation process because angular momentum is better distributed during the formation of a pair of protostars when compared to the formation of a single proto star. Astronomers suspect, based on their observation of the frequency of occurrence of pre-main-sequence binary stars, that binary formation may be the primary branch in the stellar formation process.

Sometimes, the only evidence of a binary configuration is obtained through the study of Doppler shift characteristics of radiation received from what may, from Earth, appear to be a single star. A frequency shift towards blue indicates that a star is moving toward the Earth, and one towards red indicates that the star is moving away. A cycle of shifting between red and blue, then, indicates that a star is rotating around a common center of mass. A star's orbital period can be computed through knowledge of the period of this Doppler cycle.

Many ancient civilizations knew of the existence of Beta Arietis, located 60 light years away from Earth at an apparent visual magnitude of 2.66. The star pair of this spectroscopic binary system orbits around each other with a period of 107 days. The primary of the pair is an A-type star and spins rapidly, with a velocity of 73 km/s. The secondary star is much fainter, of the type F or G. *Ashwayuja* and its Devatha *Ashwini* symbolize swiftness, ease of harnessing and the ability to heal. The *Ashwini* are called the universal messengers; astronomers use binary and multi star systems to calculate crucial[70] information in mapping the Milky Way galaxy.

Figure 45 illustrates the collapse of a molecular cloud and splitting into two with original angular momentum transferred three.

Eridanus Supervoid

Astronomical phenomenon – Black Matter, Black Energy, CMB
Nakshatra Characteristics – release, darkness, terminate, carry away

The cosmic microwave background (CMB) is the oldest light in the universe, noted in 1964 by its discoverers Arno Penzias and Robert Wilson to be "imprinted on the sky when the Universe was just 380,000 years old." According to the Big Bang theory, an early stage in the development of the

[70] Doppler shift studies are used to determine distance among stars and also the mass of individuals stars in a system of stars

universe created CMB, detectable only through sensitive microwave detectors as a faint background glow. It is present in near homogeneity in every direction, independent of stars, galaxies, or other astronomical objects.

Some regions are unusually cool relative to the average radiation levels found in most other places. A massive CMB cold spot within the Eridanus constellation spans about 5° in the night sky, centered at a longitude of 03h 15m 05s and a latitude of −19° 35' 02". One possible explanation is that the region is a "Super Void" containing no star material – one that lies between the Milky Way and the primordial CMB.

Voids are cooler than their surrounding sightlines due to the *late-time integrated Sachs-Wolfe effect*, related to Doppler red. Dark energy stretching the void as photons pass through it makes the effect more pronounced. A controversial new theory proposes that the void could be the imprint of another universe beyond our own.

Astronomers attribute various peculiar characteristics of the dwarf galaxy NGC 1156 to a dark galaxy in its vicinity that is neither directly observable nor has star formation activities. Its visible neighbor is irregular, has a larger than average core, and contains zones of contra-rotating gas. The expansionary, visible aspects of the universe can be attributed to energy from the Sun and stars. Meanwhile, the inert and dark aspects of the universe are merely reflective of the absence of such energy.

In Vedic texts, carrying consciousness away from the world of the living is the role of *Yama*. The Nakshatra *Bharanii* and its devata *Yama* represent a bridge between the lively and the inert aspects of existence. That the Eridanus Supervoid and the dark galaxy neighboring NGC 1156 are located within the Nakshatra *Bharani* is a reminder of the contradictory nature inherent to the universe.

Figure 46 highlights a region of CMB emission around a cold spot.

Conclusion

The cognition of the Vedic Rishis and their subsequent description of the vastness of space and time square exceedingly well with the snapshot of the universe modern cosmologists continue to construct today. Through these similarities, we may discern more truths from the ostensibly cryptic expression and language of the Vedas.

The Rishis weaved their rich description of the universe through the employ of three fundamental concepts: time that is cyclical rather than linear, space that is both infinitely large and infinitely small rather than defined, and a mind that is independent of both. One can only speculate how the human brain arrived at these conclusions, cognized the division of space as Nakshatras, or developed a cyclical model of time dating the present age of the current universe in the billions of years, a figure modern science has arrived at in only the past few decades.

As more people become open to the concept of consciousness existing separately from matter, the process of cognition may gain in clarity. But until that time, the accuracy of Vedic astronomical principles and the unequivocally grand scales on which the Vedic Rishis observed, described, and experienced the universe will remain at once inspiring and mysterious.

9. *References*

Abhyankar, K. (1991). Misidentification of some Indian Nakshatras. *Indian Journal of the History of Science.*

Abhyankar, K. (1996). Kaliyuga, Saptarsis, Yudhisthira and Laukika Eras. *Indian Journal of Histroy of Science.*

Achar, N. (1997). Five Year Yoga Cycle for the Performance of Vedic Rituals. *Electronic Journal of Vedic Studies.*

Burgess, E. (1998). *Surya-Siddhanta: A Text Book of Hindu Astronomy (1858).* Kessinger Publishing.

C. L. Bennett, e. a. (n.d.). First Year Wilkinson Anisotropy Probe (WMAP) Observations: Preliminary Maps and Basic Results. *Astrophysics Journal, http://arxiv.org.*

Chicago, U. o. (2013, Aug 30). Ultracold Big Bang experiment simulates evolution of early universe. *Astronomy.*

Clavin, W. (2006). *NASA's Spitzer Digs Up Troves of Possible Solar Systems in Orion.* Jet Propulsion Laboratory.

Colebrooke, H. T. (1817). Algebra, with Arithmetic and mensuration, from the Sanscrit of Brahmegupta and Bháscara. London J. Murray.

de La Fuente, M. (1998). Dynamical Evolution of Open Star Clusters. *Publications of the Astronomical Society of the Pacific.*

Dumont, P.-E. (1954). The Istis to the Nakshatras. *Proceedings of the American Philosophical Society 98.*

Gangooly, P. (1989). *The Surya Siddhanta: A Textbook of Hindu Astronomy.* Motilal Banarsidas.

Holay, P. (n.d.). The Distinctive Features of Rik-Jyotisha. *Bulletin of the Astronomical Journal of India, Vol. 26,*, 51.

JAMES, R. D. (1984). Periodic mass extinctions and the Sun's oscillation about the galactic plane. *Nature.*

Keith, A. B. (1920). *Rigveda Brahmanas: the Aitareya and Kausītaki Brāhmanas of the Rigveda (1920).* Cambridge, Mass., Harvard University Press.

Kenji Hamaguchi, M. F. (2005). Discovery of Extremely Embedded X-Ray Sources in the R Coronae Australis Star-forming Core. *The Astrophysical Journal.*

Lancelot Wilkinson, B. (1891). *Translation of the Surya Siddhanta.* Baptist Mission Press.

Madhavananda, S. (1950). Brihadaranyaka Upanishad - Shankara Bhashya. *Internet archives http://archive.org/.*

Minkowski, C. (n.d.). *Competing Cosmologies and the problem of Contradictions in Sanskrit Knowledge Systems.* Princeton University.

Naik, P. (n.d.). Samanta Chandra Sekhar. *Bulletin of the Astronomical Society of India, Vol. 26*, 33.

Price, J. (n.d.). *Applied Geometry of the Sulba Sutras.* Sydney: School of Mathematics, University of New South Wales.

Ravishankar, S. S. (2002). *Time : Exploring the various aspects of time through the ages.* Vyakti Vikas Kendra India, Publications Division.

Ray, B. (2009). *Different Types of History.* Pearson Longman.

Saraswati, S. D. (2011). *Satyarth Prakash, 2nd Ed.* Adhyatmik Shodh Sansthan-Subodh Pocket Books.

Schaffer, S. (1980). Newton on the Ganges: Asiatic Enlightenment of British Astronomy. *http://www.youtube.com/watch?v=YeKWqoiMo3Y.* Stanford Humanities Event.

Sidharth, B. (1998). The Calendric Astronomy of the Vedas. *Bulletiin of the Astronomical Society of India, Vol 26.*

Smith, G. (n.d.). Six Thousand Year Barrier. *http://web.nickshanks.com/history/sixthousandyears.*

Sri Aurobindo, A. G. (1971). *The secret of the vedas.* Sri Aurobindo Ashram Publication Department.

Steinmetz, M. (n.d.). *Formation of spiral and elliptical.* Munchen: Astronomy and Astrophysics Letters.

Stoyan, R. (n.d.). *Atlas of the Messier Objects: Highlights of the Deep Sky.* Cambridge University Press.

Swarup, L. (n.d.). *Nighantu and the Nirukta.* Motilal Banarasidass.

Upadhyaya, A. (1998-08). *Siddhanta Darpana.* Nag Publishers.

van der Waerden, B. L. (1980). Two Treatises on Indian Astronomy. *Journal for the History of Astronomy, Vol. 11*, 50.

Virendra Nath Sharma, J. (n.d.). *Sawai Jai Singh and His Astronomy.* Motilal Banarsidass.

Waller, B. W. (2013). *The Milky Way: An Insider's Guide.* Princeton University Press.

Wilke, A. (2011). *Sound and Communication: An Aesthetic Cultural History of Sanskrit Hinduism.* Walter de Gruyter.

Winter, L. M. (2008). *Extragalactic X-ray Surveys of ULXs and AGNs.* ProQuest.

Zinner, E. (2003). An isotopic view of the early solar system. *Science.*

www.ingramcontent.com/pod-product-compliance
Lightning Source LLC
Chambersburg PA
CBHW072026190526
45166CB00015B/508
* 9 7 8 1 4 9 3 5 1 0 3 1 3 *